THE ASTROLOGER'S ASTRONOMICAL HANDBOOK

THE ASTROLOGER'S ASTRONOMICAL HANDBOOK

by

JEFF MAYO, D.F.Astrol.S.

(Tutor, Faculty of Astrological Studies)

Author of *Teach Yourself Astrology*

Illustrations by the Author

L. N. FOWLER & CO. LTD.
15 NEW BRIDGE STREET
LONDON EC4 V6BB

SBN 8524 3058 2

Printed offset in Great Britain by
The Camelot Press Ltd., London and Southampton

Contents

Foreword

Astronomy is a science concerned with observing, recording, accounting for and predicting the motions of the celestial bodies, and determining their physical elements.

Without the calculated positions of the Sun, Moon, and planets provided by astronomers, astrology would not exist as we know it today. Astronomy is the basis of astrology. It provides the raw materials, the tools and equipment, and, indeed, for the serious and progressive-thinking astrologer, the inspiration for new theories of interpretation.

It is not enough for the would-be astrologer to learn that the planets and signs are symbols, needing only to be interpreted in terms of character traits or possible events as indicated in text-books and handed down by tradition. Old and accepted theories need periodical checking, revising, in the light of fresh knowledge and deeper understanding. And, above all, the student needs to be conversant with the astronomical framework upon which the whole structure of his astrological interpretation is based.

This volume is offered as an informative reference book and as a guide towards a clearer understanding of the derivation and elements of the basic factors which make up the astrological chart. Certain of the contents may not seem to have a direct bearing on astrology, and yet all that is within these pages should help to convey to the reader some measure of wider understanding of the complex interplay of forces which create and re-create the cosmic patterns, within which our own planet is an exciting and essential feature.

THE GEOCENTRIC FRAMEWORK

1   *The Basic Framework*

This book, like any other book, must have an apt starting-point, and what better than to begin with the Earth we live on.

From our standpoint of Earth we look out at the rest of the universe. The Sun and Moon, the planets, and even the faraway constellations, *appear* to revolve about us. Except, in a sense, for the Moon this is not really what is happening. The Sun, and not the Earth, is of course the common centre of gravity for all the planets in the Solar System.

However, it is perfectly correct for the astrologer to symbolize the Earth as a kind of axis round which the rest of the cosmic bodies revolve, because the astrologer is concerned with the *angular relationship* of the Sun, Moon, and planets *as seen from the Earth*.

Hence, the small circle in the centre of the astrological chart (Fig. 1) symbolizes the Earth as the central point of reference.

The positions of the planets given in the ephemerides, from which the astrologer can calculate the planetary pattern as seen from the Earth for any given moment, are the *geocentric* measurements (Gk: *ge*, Earth). If the Sun were the central point of reference the measurements of the planets thereto would be termed *heliocentric* (Gk: *helios*, Sun).

To enable the astronomers to calculate the motions and angular positions of the planets and other celestial bodies relative to the Earth they need a basic framework and points of reference. This is our next step, to understand the

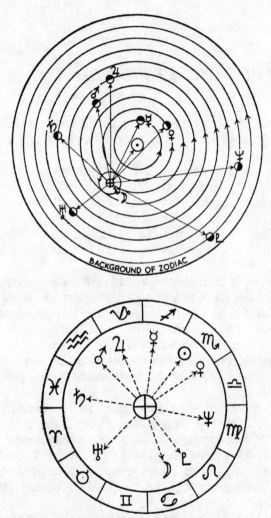

Fig. 1. (*Bottom*) Positions of the planets relative to the Earth symbolized in birth-chart. (*Top*) Same positions of the planets in their orbits relative to the Earth

"geocentric framework" employed by the astronomers, with particular emphasis on those circles and their points of intersection used by astrologers.

The Celestial Sphere

When we look out at the stars it is easy to imagine they are points of light speckling the inside of a spherical dome —the visible top half of a whole sphere, the lower half being beneath the level of our feet—and that we are at its centre. Thus, we speak of the *Celestial Sphere*, and if you refer to Fig. 2 you will see a number of circles contained within an outer circle. This latter we call *the Meridian*.

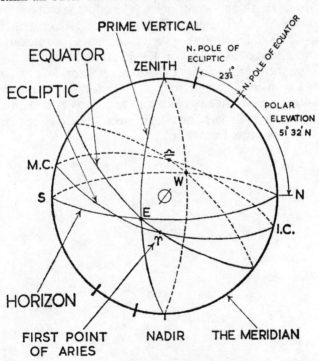

Fig. 2. The Celestial Sphere
(Described for Polar Elevation 51° 32′ N, the latitude for London)

Any circle, the plane (or level) of which passes through the centre of the Earth, is a *great* circle. Referring to Fig. 2 with the tiny symbol of the Earth and its north–south axis at the centre of the diagram, you should note that every circle shown is a great circle. You may understand this more clearly if you think of any one circle not as a hoop (which it might here resemble) but as a flat disc. We have to distinguish between a *great* circle and a *small* circle. A small circle is any circle the plane of which does NOT pass through the centre of the Earth.

For example, the *equator* is a great circle. Its plane intersects the Earth's centre and is, therefore, equidistant from its north and south poles. The equator corresponds to latitude 0°. All other parallels of latitude, which with the meridians of longitude are the co-ordinates for measuring the position of any place on Earth, are *small* circles.

The three great circles of *horizon*, *equator*, and *ecliptic* are the main circles of reference for locating a planet's position relative to any place on Earth. It is essential that the astrologer understands their interrelationship and role in the geocentric framework.

2   The Horizon System

The Horizon, Zenith, Nadir

The *horizon* is the great circle shown in Fig. 3 which you can trace by the four cardinal points from N to E to S to W and back to N. This creat circle is called the *rational* or *true* horizon. Its poles, the extremities of its axis which is

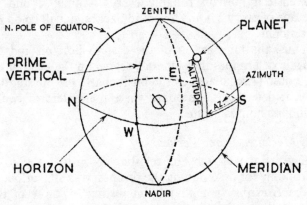

Fig. 3.   Horizon System of Celestial Co-ordinates

perpendicular to its plane, are the *zenith* (directly overhead for the observer) and the *nadir* (directly beneath the observer). A line drawn between zenith and nadir would be in the direction in which gravity acts—as with a plumb-line.

When the great circle of the rational horizon is produced to meet the heavens (i.e. extended beyond the sphere of the Earth) it is called the *celestial* or *sensible* horizon. The rational horizon is not to be confused with the circular line formed by the apparent meeting of the earth and the sky, which is a *small* circle, is called the *visible* or *apparent* horizon, and is parallel to the rational horizon.

The very important function of the horizon with reference to the astrological chart is that its *eastern* point of intersection with the ecliptic determines the *Rising Degree*, or *Ascendant*, for any given time and place.

The horizon for any specified place on Earth remains constant, just as the zenith at right-angles to the horizon will be that imaginary point constantly overhead the observer. But as the Earth continuously rotates on its axis a different degree area of the ecliptic will rise above the eastern horizon roughly every 4 minutes, and an average of every 2 hours a new zodiacal sign will be on the Ascendant.

Thus, the intersection of the eastern horizon and the ecliptic determines the sign and degree on the cusp of the 1st house (Ascendant) of the birth-chart; or, put another way, it determines the point where the system of twelve mundane houses begins.

The Meridian, Prime Vertical

One of the systems of celestial co-ordinates used in astronomy for determining the position of a star or planet is the *Horizon System*. It is interesting to know of this system, though its co-ordinates of *altitude* and *azimuth* are not employed by the astrologer. However, if you refer to Fig. 3 you will see the framework of reference circles used in the Horizon System. These are the *horizon, the Meridian*, and the *Prime Vertical*.

The Meridian is often referred to as the north–south great circle. This is because it passes through the zenith and nadir (poles of the horizon), the north and south poles of the equator, and the north and south points of the horizon. This great circle is called *the* Meridian to distinguish it from all other meridian circles. The Meridian is an integral part of each of the three systems of celestial co-ordinates:

a. It is a *vertical circle* in the Horizon System (Fig. 3)
b. It is a *meridian of right ascension* in the Equatorial System (Fig. 4)
c. It is a *meridian of longitude* in the Ecliptic System (Fig. 5)

If we look again at Fig. 2 another reason for the distinction of being termed *the* Meridian becomes obvious: all other circles are contained within its circle. Yet another significant point is that the Sun crosses the Meridian at midday (i.e. when it is midday *anywhere* on Earth). At midday the ecliptic (Sun's apparent path) intersects the great circle of the Meridian, and this point of intersection is known to astrologers as the *Midheaven* (M.C., from Latin: *Medium Coeli*), or degree of culmination.

The *Prime Vertical*, sometimes called the east–west great circle, is a vertical circle passing through the zenith and nadir, and east and west points of the horizon (Fig. 3). All other vertical circles (called *verticals*) which also pass through the zenith and nadir, unlike the Prime Vertical, do *not* pass through the east and west points of the horizon. Thus, all other verticals are referred to as *secondaries* to the horizon. In passing through the east point of the horizon it is the first, or *prime* vertical. Its plane corresponds to the points of intersection of the horizon and the equator.

The horizon, equator, and Prime Vertical (great circles linked by a common factor, that of passing through E–W —see Fig. 2) can be spoken of as secondaries to the Meridian—because the poles of the Meridian are the east and west points of the horizon.

It is as well that the astrologer should understand the various astronomical terms, so briefly we shall define *altitude* and *azimuth*, the two co-ordinates of the Horizon

System. The position of a planet when described by altitude is measured in degrees, minutes, and seconds from the horizon up towards the zenith. It is the planet's angular distance from the horizon, measured along a vertical. Azimuth is its measurement along the horizon in degrees, minutes, and seconds *westward* from where the Meridian cuts the south point of the horizon, to the vertical containing the planet; or, expressed as the angle which the vertical through the planet makes with the Meridian. Astronomers speak of an object's (star, planet) *zenith distance*, which is its angular distance from the zenith downwards towards the horizon—the complement of the planet's altitude.

Polar Elevation (Fig. 2) or *polar altitude* is the height of the pole above the horizon at a given place, and is equal to that place's latitude or angular distance from the equator.

3 *The Equatorial System*

The Equator, First Point of Aries

The *terrestrial equator* is a great circle corresponding to the Earth's largest circumference, midway between its north and south poles around which the galaxies of stars appear to revolve.

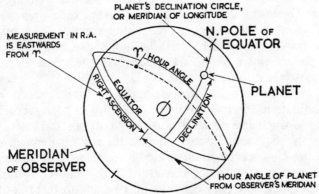

Fig. 4. Equatorial System of Celestial Co-ordinates

The *terrestrial poles* are the two points where the axis of rotation meets the Earth's surface. The *celestial poles* are the terrestrial poles extended to infinity.

If we imagine the plane of the terrestrial equator projected beyond the Earth, this "extended" plane is called the *celestial equator*. Why should there be two names for what is, in effect, the same great circle? Simply because the plane of the equator is used as a circle of reference for fixing the position in the sky of a star or planet as well as for locating the situation of any place on Earth.

The terrestrial equator is a great circle known as a *parallel of latitude* corresponding to latitude 0°. On the terrestrial sphere, parallels of latitude and *meridians of longitude* are the well-known co-ordinates used for finding a place on a map. New York, for instance, is positioned on the parallel of latitude which is 40° 45′ north of the great circle of the equator. But the parallel of latitude on which New York lies is, like all other latitudes north or south of the equator, a *small* circle. Their planes do *not* pass through the centre of the Earth.

The astronomer, with reference to the circle of the celestial equator, uses similar co-ordinates for fixing the position of a heavenly body. These co-ordinates belong to the *Equatorial System* (Fig. 4), and are called *parallels of declination* (which correspond exactly to similar measurements in terrestrial latitude), and *meridians of right ascension*. A planet exactly on the plane of the equator has no declination (0°).

Measurement in degrees, minutes, and seconds of declination is in the direction of either of its poles, according to whether a planet is north or south of the zero point of the equator. Because the angular measurement of declination corresponds exactly to that of terrestrial latitude you must not confuse it with *celestial latitude*, which is the angular distance of a planet north or south of the ecliptic, and refers to the Ecliptic System which will be dealt with presently.

For either terrestrial longitude or right ascension there is no natural zero, so an arbitrary choice is made. The great circle passing through the poles of the equator and the position of Greenwich (England) is where measurement in terrestrial longitude begins. We speak, for example, of New York being 74° west, which means it is situated 74° west of the Greenwich meridian. The starting point for measurement along the celestial equator in right ascension is where the ecliptic and the equator intersect at the vernal equinox, and is called the *First Point of Aries*.

There are two intersections of the ecliptic and equator, the second being the *First Point of Libra*, occurring at the autumnal equinox. Both intersections are spoken of as the *Equinoctial Points*. The word "equator" derives from its other name, the *equinoctial*, because at the equinoxes (Equinoctial Points) day and night are of equal length (Latin: *nox*, night). The vernal equinox or First Point of Aries is denoted by the symbol ♈, as used in astrology for the zodiacal sign of Aries. The Sun is then crossing from south to north of the equator. The autumnal equinox or First Point of Libra is denoted by the symbol ♎, as used for the zodiacal sign of Libra, and it is when the Sun crosses the equator again, this time from north to south.

The First Point of Aries is very important in astronomy. Just as the vernal equinox, with which it corresponds, is an important phase of the year for those of us living in the Northern Hemisphere. For then our spring begins. In ancient times the beginning of the *year* was reckoned from this moment when the Sun crossed the equator and began rising higher each day in the heavens. Due to *Precession* (p. 76) the constellation (not zodiacal sign) which lies behind the Sun at the vernal equinox changes gradually through the centuries. However, the First Point of Aries retains its name, and for the astrologer this is the moment when the cycle of the twelve Signs of the Zodiac begins with 0° of the sign Aries.

The Equatorial System is defined by the *direction of the Earth's axis*. Why? Because due to the Earth's rotating motion the complete circle of the ecliptic is brought into view of any part of the Earth's surface within the approximate period of 24 hours. Although the Sun is only *exactly* on that point of intersection of the equator and ecliptic twice in one year (around 21st March and 23rd September), these intersections of First Point of Aries and First Point of Libra are reference points which always exist. The First Point of Aries is a constant point of reference since once in every 24 hours it will cross the meridian of any place.

We have already seen how a planet's position by the Equatorial System co-ordinate of declination is measured. How do we finally fix the planet's location in the heavens by the other co-ordinate, *right ascension*? First, let us define right ascension.

We know there are two co-ordinates for locating the position of a place on Earth, just as there are two co-ordinates for locating the position of a planet in the heavens, relative to the Earth. The former, terrestrial or geographical co-ordinates, are latitude and longitude. The latter, celestial co-ordinates, are declination and right ascension. Terrestrial latitude corresponds to declination: both are measured north or south *from* the equator. Terrestrial longitude corresponds to right ascension: both are measured *along* the equator.

The Meridian of Greenwich marks the beginning of measurement in terrestrial longitude, in a westwards direction for 180°, or in an eastwards direction for 180°. The First Point of Aries might be thought of as the "Greenwich" of the celestial sphere, for it is here that measurement in right ascension begins. But right ascension is measured in one direction only: *eastwards* (Fig. 4). It is measurement taken in degrees, minutes, and seconds of arc, from 0° through a full circle of 360°; or in hours, minutes, and seconds of time, from 0 hours through a whole cycle of 24 hours. *Hour angle* is a measurement westwards from the First Point of Aries, which is discussed on page 42.

Thus, right ascension can be defined as *measurement eastwards along the equator in degrees of arc, or in hours of time, from the First Point of Aries.*

We have seen that the diurnal motion or rotation of the Earth on its axis causes the Sun and planets to appear to go round the Earth from *east to west*, whereas in reality they, like the Earth, move from *west to east*. We see, therefore, that measurement in right ascension is in that direction which the Earth rotates, eastwards. More significantly, it is the direction in which the planets (including Earth) revolve round the Sun. By the measurement in right ascension we can each successive day determine the angular distance a planet is *eastwards* from the First Point of Aries, as seen from our standpoint of Earth.

The full circle of the equator equals 360° of arc, or 24 hours in time, in terms of right ascension. When we speak of the Sun's transit of 1 hour in right ascension (R.A.) corresponding to 15° of arc, we mean that due to the rotation of the Earth the Sun in 1 hour appears to pass over an area of the Earth's surface equal to R.A. 15°. This will, perhaps, be better understood if we think of 1 hour as one-twenty-fourth part of the 24-hour cycle of one day; coincident with 15° of arc representing one-twenty-fourth part of a full circle of 360°.

Thus, if 15° of arc equals 1 hour (60 minutes) of time, 1° of arc will equal 4 minutes of time. Right ascension, remember, corresponds to terrestrial longitude. It follows, therefore, that the Sun will take roughly 4 minutes to pass over an arc of 1° of the Earth's surface. Amsterdam in the Netherlands is 5° east of the Greenwich meridian. As the Sun appears to move from east to west, sunrise in Amsterdam will occur 20 minutes (5° × 4 minutes = 20 minutes) earlier than it will in London.

Although measurement in right ascension is not normally used by the astrologer it is as well that he should understand this term, since from this co-ordinate the positions of the Sun, Moon, and planets are converted into *celestial longitude* as they appear in the yearly ephemerides.

4 *The Ecliptic System*

The Ecliptic, the Zodiac

We speak of the *ecliptic* (Fig. 5) being the apparent path of the Sun in the heavens, the great circle the Sun appears to follow in its journey round the Earth taking a year to complete one cycle. In reality it is the Earth that orbits

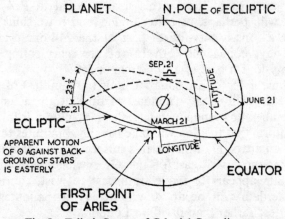

Fig. 5. Ecliptic System of Celestial Co-ordinates

the Sun, the Sun being the Earth's centre of gravity in common with the other planets in the Solar System.

It is not entirely true to say that the ecliptic is, therefore, an *imaginary* great circle traced in the heavens, the plane of which passes through the centre of the Earth. The ecliptic, as well as being the actual path followed by the Earth, also corresponds very closely to the plane of the orbits about the Sun traced by the other major planets.

From the earliest records of man's study of the celestial sphere it is known that this apparent course of the Sun and planets in the sky was recognized, and plotted against the background of constellations of "fixed stars"—so called because due to their immense distances from the Earth they scarcely shift their positions relative to the Earth over several centuries, and appear fixed in space. Certain patterns of grouped stars, or constellations, "through which the Sun and planets moved", were associated by the earliest of priest-astrologers with mythological figures significant to their primitive stellar religion. Thus, over the centuries these very special constellations came to be known as the *Signs of the Zodiac*. Zodiac is derived from the Greek *zōon*, meaning a living thing. Only one of the twelve signs (Libra, the scales) does not represent a living creature. Hence, the zodiac is sometimes called the "Circle of the Beasts". The twelve constellations have only a mythological connection with the twelve signs.

The *zodiac* can be defined as a belt of the sky which has the ecliptic as its centre. This belt is considered to extend about 8° either side (north and south) of the ecliptic, a total of 16°, within which the Sun, Moon, and major planets always remain. Pluto is the exception, its inclination to the ecliptic reaching as much as 17° as measured in *celestial latitude*.

The ecliptic is so called because *eclipses* can only occur when the Moon is in or very near to it. The Sun can never be away from the ecliptic, since this is a great circle the plane of which passes through the centre of the Sun as well as through the centre of the Earth.

As we have already said, the *vernal equinox* is when the Sun's path along the ecliptic crosses the equator from south to north. This occurs around 21st March (Fig. 5), and in the Northern Hemisphere is heralded as the first day of spring. Each successive day the Sun by noon appears higher in the heavens. This is because the angle of the Earth's equator to the plane of the ecliptic increases.

This angle is measured in *declination*, and when the Sun has reached its maximum declination north of the equator we say it is at "declination $23\frac{1}{2}°$ north", which occurs in mid-summer in the Northern Hemisphere, around 21st June. This tells us that the angle between the plane of the equator and the plane of the ecliptic, measured along a declination circle northwards from the equator, will be $23\frac{1}{2}°$. Or, put another way, the pole of the ecliptic is about $23\frac{1}{2}°$ from the pole of the equator (Fig. 2). The Sun's maximum declination north marks the *summer solstice*. It is then that increase in declination is halted, the Sun, as it were, stands still (Latin: *sistere*, to cause to stand still), and then declination starts to decrease, until about 21st September the plane of the equator and the centre of the Sun are in line. We say that the Sun has then declination 0°, zero, as it crosses from north to south of the equator. This marks the *autumnal equinox*, when day and night throughout the world are again of equal length—as they were at the vernal equinox.

As the Sun's declination southwards increases, so in the Northern Hemisphere the Sun's maximum altitude at noon above the horizon gets lower, until around 21st December the Sun reaches maximum declination $23\frac{1}{2}°$ south, which marks the *winter solstice*.

The *Ecliptic System* (Fig. 5) is defined by the *plane of the Earth's orbit*, which fixes the position of the ecliptic. In simpler terms we would say that the plane of the Earth's orbit round the Sun determines where the ecliptic lies in relation to the Earth's equator. To plot the position of a planet by this system we use the co-ordinates of *celestial longitude* and *celestial latitude*.

Celestial longitude is that co-ordinate of prime importance to astrologers. It is measurement in degrees and minutes of arc eastward from the First Point of Aries *to the meridian of longitude which passes through both the planet and the poles of the ecliptic.* When, for example, we say that Venus is in Aries 19° we know that Venus is a distance of 19° eastward from where the ecliptic intersects the equator. Note that celestial longitude is measured in the same direction as right ascension, i.e. eastward.

Celestial latitude is measurement in degrees and minutes north or south of the ecliptic (Fig. 5), though it is rarely applied by astrologers for correcting a planet's position plotted by longitude.

The diurnal motion of the Earth (rotation on its axis) does not affect the co-ordinates of latitude and longitude. These refer to the ecliptic (plane of Earth's orbit round the Sun).

TIME FACTORS

Calculation of an astrological chart, whether it be the birth of a human child or the formation of a business company, requires the *date*, *place*, and *time* of the event's occurrence. The more accurate the known time, the greater the opportunity for an accurate interpretation of the life-potentialities by the astrologer charting the event.

The two basic factors of *time* and *place* (space) are part and parcel of the astronomer's equipment also. In the previous chapters you have been shown the framework of circles and reference points derived from the Earth's axial rotation, orbital revolution, and angular relationship to the Sun, employed by the astronomer for plotting the exact location of a celestial body.

It is not necessary that the astrologer remembers the exact details and figures concerning the various time systems, but he is a wiser and better astrologer who takes the trouble to understand these basic factors.

It is common knowledge to every schoolchild that man reckons the age of himself or other growth-formations on Earth in four units of time: in terms of a year, a month, a week, or a day. These are but four basic measurements in time. For astronomical purposes there are other time systems also employed.

5 The Year

There are several kinds of year. Though I shall briefly mention some of these you need only concern yourself with the two of special importance: the *tropical year* and the *sidereal year*.

The *tropical year* (Gk: *trope*, turning) is the time that elapses between successive passages of the Sun through the vernal equinox, or First Point of Aries. That is, the moment the Sun crosses the Earth's equator from south to north measurement *in time* of the cycle of the tropical year begins; it ends the moment the Sun crosses the equator again from south to north. In reality, of course, this is the exact period in which the Earth has gone round the Sun between two successive equinoxes, and as it did so the changing angle between the inclination of the Earth's axis and the ecliptic made the Sun appear to move from south to north of the Earth's equator. This period measures 365 days 5 hours 48 minutes 46 seconds in mean solar time. Because this year or period is defined by successive passages of the Sun through the vernal equinox it is sometimes referred to as the *equinoctial year*.

The tropical year has also been called the *astronomical year* because of its usage in astronomy. Another name for it is the *natural year*, since it marks a natural unit of time for man. Likewise when it is also called the *seasonal year* or the *year of the seasons* this is because the seasons as we know them derive from this Earth–Sun cycle. Indeed, time would not exist for us if all the bodies in the Solar System remained motionless in space. Man was able to invent this convenient form of measurement because time depends upon the Earth's axial rotation (time unit of the day) and its revolution of the Sun (the year).

The *sidereal year* (Latin: *sidus*, star) is measured *relative to the stars*. Hence it has been known as the *astral year*. The stars provide a convenient and essentially fixed system of reference for man. Therefore, the interval between the passage of the Sun across a secondary to the ecliptic which passes through some fixed star and its return to that same star is called a sidereal year. This equals 365 days 6 hours 9 minutes 9·5 seconds in mean solar time. This actually is the true year, the time taken by the Earth to complete a revolution round the Sun from a given direction in space to the same direction again. Or we might say that in a period of one sidereal year the Sun will have returned to exactly the same position against the background of constellations as seen from the Earth.

The sidereal year is just over 20 minutes longer than the tropical year. Why is this? If the First Point of Aries were a fixed point among the stars the sidereal and tropical years would be of exactly equal length, but it is not a fixed point due to the Precession of the Equinoxes (p. 76). This 20 minutes difference when expressed in celestial longitude equals roughly 50·26 seconds, which is the annual displacement or retrograde motion of the First Point of Aries.

For chronological and civil purposes neither the tropical nor the sidereal years would be of much practical use to man in terms of a calendar. The *calendar year* must have an exact number of days, and it must mark the recurrence of the seasons. The tropical year marks the recurrence of the seasons, but its length is not an exact number of days. So the *civil year*, or calendar year, was invented.

The civil year has 365 days, and 366 days in each *leap year*. It was Julius Caesar who reformed the Roman calendar by introducing a leap year every fourth year. Hence, in 45 B.C. the *Julian Calendar* was established.

Leap years were chosen to be those years the number of which is divisible by 4, such as 1964. Three ordinary years and one leap year exceed four tropical years by just under 45 minutes. This difference amounts to 3 days 2 hours 53 minutes 30 seconds in 400 years. Consequently the vernal equinox occurs continually earlier in the Julian Calendar, and in 1582 it fell on 11th March instead of on the 21st. To remedy this the calendar now in use, the *Gregorian Calendar*, was introduced by Pope Gregory XIII in 1582. It was decreed that 3 days shall be omitted in every 400 years. Thus, every year whose number is a multiple of 100 will be an ordinary year of 365 days instead of a leap year, unless the number of the century is divisible by 4, in which case it will be a leap year. For example, 1700, 1800, 1900, and 2100 are not leap years, but 1600 and 2000 are. When the Gregorian Calendar finally superseded the Julian Calendar in England in 1752 it was also decreed that the year begin on 1st January instead of 25th March. But to this day the financial year is still reckoned from 6th April (being the new calendar equivalent of 25th March—the 11 days difference that the Gregorian Calendar is ahead of the Julian Calendar).

The period of the Sun's revolution relative to the apse line (p. 85) is the *anomalistic year*. It is the interval between successive passages through perigee or apogee, equal to 365 days 6 hours 13 minutes 53·01 seconds. The Sun's revolution around the centre of our galaxy (p. 52) is called the *cosmic year*, equalling about 200 million years.

6 *The Month, the Week*

The Month

The *month* refers to the Earth–Moon cycle, and its name derives from the Moon.

Since the earliest recorded use of a calendar there have been numerous attempts to combine all three factors of Earth rotation, Sun, and Moon, but without success. The awkward factor is the Moon. Its orbital period of the Earth, which combined with the Sun produces the well-known four phases or quarters, just does not tie in neatly with the day (Earth rotation) and year (Earth–Sun cycle). Many countries in ancient history made their months begin with the New Moon, which in those times was taken to be when the Moon's thin crescent could first be seen in the west after sunset. But as the months were expected to have an exact number of days it was arranged to make them alternately 29 and 30 days long. This meant the year was 354 days. The Mohammedan Calendar uses the *lunar month* to this day. Some nations in the past kept their lunar-based years reasonably in step with the solar year by inserting a thirteenth month roughly every three years.

So far as astrology is concerned the student will hear three kinds of months referred to, and it is as well that he becomes familiar with their derivation. These are the *calendar month*, the *synodic month*, and the *sidereal month*.

The *calendar month* is, of course, the familiar one-twelfth of our year, January, February, and so on. This month is purely a convenient division of the year, and is not based on the actual motions of the Moon. The months are not all the same length, for as the ditty says:

Thirty days hath September,
April, June, and November;
All the rest have thirty-one
Excepting February alone,
Which hath but twenty-eight days clear,
And twenty-nine in each leap year.

The average length of the calendar month is slightly in excess of the *synodic month*. A *synodic period* is time taken by a planet or satellite to return to the same relative position with respect to the Sun and Earth. Thus, the synodic month is the interval between consecutive conjunctions of the Sun–Moon as seen from the Earth. Astrologers generally refer to it as a *lunation*, the period from New Moon to New Moon. The mean or average synodic month is 29 days 12 hours 44 minutes 2·7 seconds. The length of the synodic month varies by as much as 13 hours, due mainly to the eccentricity of the Moon's orbit of the Earth. Synodic derives from the Greek *súnodos*, Latin *coitus*, the ancients likening the Sun–Moon conjunction to the copulation of man and woman.

The *sidereal month* is the Moon's period of revolution relative to the stars, which is more than two days shorter than the synodic month (Fig. 13). The mean length of this period is 27 days 7 hours 43 minutes 11·5 seconds. As seen from the Earth this is the average time the Moon takes from a given star back to the same star. Due to the many disturbances (perturbations, p. 76) the Moon's motion undergoes, the sidereal month can vary by as much as 7 hours.

By a simple equation the Moon's sidereal period, the sidereal year, and the synodic month can be found to be curiously related. That is, the number of sidereal months in a sidereal year is *exactly* one greater than the number of synodic months, being 13·369 + and 12·369 + respectively.

There is yet another month known to astronomers. The

draconitic or *nodical month*. This refers to the period taken by the Moon to pass round its orbit from one node (p. 81) back to the same node. Since the Moon's nodes move backward along the ecliptic, similar in effect to the Precession of the Equinoxes, the nodical month is not the same as the time taken from a given star back to the same star. In fact, the nodical month (27·21222 days) differs from the sidereal month (27·32166 days) in the same way as the tropical year differs from the sidereal year. The nodical month is an exact submultiple of the interval at which eclipses can occur. Its connection with eclipses has an interesting link through its other name *draconitic*, which originated in the time when earlier astrologers believed that a dragon swallowed the Sun at a total eclipse.

The *anomalistic month* is the interval between successive passages of the Moon through perigee (p. 85), which averages 27 days 13 hours 18 minutes 37·5 seconds of mean solar time.

The Week

The *week* is the third unit of time used in the calendar. Its length corresponds roughly to one of the four phases or quarters of the Sun–Moon cycle or lunation. Its origin goes back nearly 3,000 years before the Christian era to the great period of early astronomy of the Babylonians. They devised the 7-day week.

Even the names of the weekdays associated with the planetary gods are retained to this day. Sun-day, which marks the beginning of the week, is significantly named after the central body and source of light in the Solar System (German *Sonntag*). Mon-day is the Moon's day, closely resembled in the English name, the Anglo-Saxon *Monandaeg*, and the German *Montag*. The Italian *Lunedi*, and French *Lundi*, derive from lunar. Tues-day is Mars' day. The Italian *Martedi*, and French *Mardi*, deriving from Mars, the ancients' god of war, and the English *Tuesday*, and Anglo-Saxon *Tiwesdaeg*, relating to his Germanic counterpart, Tiw. Wednes-day is Mercury's day, clearly defined in the Italian *Mercoledi*, and the French *Mercredi*. Again the English word and the Anglo-Saxon *Wodneesdaeg* relate to his Germanic counterpart Woden. Thurs-day is Jupiter's day. The English name and the Anglo-Saxon *Thuresdaeg* are associated with Thor, Scandinavian god of thunder, whilst the Germans name it "thunder-day", *Donnerstag*. The Italian *Giovedi*, French *Jeudi*, and Latin *Jovis dies*, derive from the Roman Jove (Jupiter). Fri-day (Fry-day) is Venus' day or *Freya's* day, as can be seen in the Italian *Venerdi*, French *Vendredi*, Latin *Veneris dies*. Her counterpart, Frigga, can be recognized in the Anglo-Saxon *Frigedaeg*, and the German *Freitag*. Satur-day is Saturn's day, associated with the English name, the French *Samedi*, and the Anglo-Saxon *Saeternesdaeg*.

7 The Day

The period of the Earth's rotation on its axis determines the fourth basic unit of time, the *day*. It may be measured with respect to the stars (*sidereal day*), the true or apparent Sun (*apparent solar day*), or the mean Sun (*mean solar day*). Our day of 24 hours, divided into 12 hours before noon (a.m., Latin: *ante meridiem*), and 12 hours after noon (p.m., Latin: *post meridiem*), comes to us from the Babylonian era of around 5,000 years ago.

The true period of the Earth's rotation is called the *sidereal day*. This is the interval between two successive transits of a fixed star over the meridian of any place. Put another way, when for example the meridian of Greenwich, due to the turning of the Earth, comes exactly in line with a certain fixed star, we can say that a new sidereal day begins at Greenwich. When the Earth has turned a full circle and the meridian of Greenwich is again exactly in line with that same star, the sidereal day is completed, and a new sidereal day begun. In actual practice the astronomers use the First Point of Aries to define the sidereal day, and the moment this Point crosses the meridian of *any given place*, corresponding to 0h. 0m. 0s. *sidereal time*, the sidereal or *true* day *for that given place* begins.

The Sun transits the meridian of any given place twice in 24 hours. When it is noon (for *any* place on Earth) the Sun is that moment crossing the *upper meridian*. At midnight it crosses the *lower meridian*, so called because at midnight for a given place the Sun is on the opposite side of the Earth and is, it would seem, "beneath one's feet"; whereas at noon the Sun is overhead, hence upper meridian. When we speak of this actual movement of the Sun around the Earth we are speaking of the *true* Sun, which is visible in daytime. Of course, as has been said before, it is the Earth turning on its axis that makes the Sun appear to be the one that is moving. Thus, because it is the *apparent* motion of the Sun, the interval between two successive passages of the Sun across the *lower* meridian we call the *apparent solar day*. Why does the apparent solar day begin when the Sun crosses the *lower* meridian? Because our day begins at midnight, and it is then when the Sun is crossing the lower meridian. The apparent solar day, like the sidereal day, is also measured through *exactly* 24 hours. But a sidereal day of 24 hours of sidereal time equals only 23h. 56m. 4·09s. of *mean solar time*. In other words, each successive day the stars rise and set about 3m. 56s. earlier due to this difference between sidereal and mean solar days. In one year the number of sidereal days exceeds that of mean solar by exactly one.

Apparent solar day must not be confused with *mean solar day*. The difference is this. The orbit of the Earth about the Sun does not describe a circle, but an ellipse. This means that the Earth moves round the Sun with variable speed (Chapter 12). At *perihelion* (early-January) the Earth is at its closest to the Sun and it also moves faster than at any other point in its orbit. At *aphelion* (early-July) the Earth is farthest from the Sun, and moving at its slowest. Therefore the *apparent* speed of the Sun as seen from the Earth varies at different times of the year. The apparent solar day from solar noon to the following solar noon is 51 seconds longer around 23rd December as it is around 23rd September. One can see that it would be rather silly to make clocks which kept pace with these variations in the apparent speed of the true Sun. So a fictitious Sun was invented and was called the *mean* Sun. Its rate of motion is the average of all the variations in speed made by the true Sun in the course of one year.

But it must be remembered that the mean Sun is simply a *point*, since it would be physically impossible for a *body* to move *uniformly* in this manner. It is a point which moves round the ecliptic and is never very far in hour angle from the true Sun, but it always moves at the same fixed pace, which the true Sun does not.

The mean solar day is measured through 24 hours in terms of *mean solar time*, and is equal to the *average* apparent solar day—which in terms of mean solar time varies in length at different periods of the year.

The *civil day* is the day beginning at midnight, and in many countries (as in the United Kingdom) this day is divided into 12 hours from midnight to noon, and 12 hours from noon to midnight. Other countries calculate the hours of the civil day from 0 to 24.

The *astronomical day* prior to the end of 1924 began at noon and was reckoned in terms of 0–24 hours. Thus 2 hours after midnight would be 14.00 hours. But since the start of 1925 the astronomical day has begun at midnight, the same as the civil day, though still counting from 0 to 24 hours.

The *lunar day* is used in calculations of high- and low-water for tidal predictions, and is the interval between two consecutive transits of the Moon across the upper meridian of a place.

8 Time Systems

Sidereal Time

We have seen that the sidereal day is the true period of the Earth's rotation, measured between two successive transits of a given star over the observer's meridian. The sidereal day is reckoned in *sidereal time*, which has been called "time reckoned by the stars". As has been said, however, the sidereal day begins in practice when the First Point of Aries, and not a given star, is on the observer's meridian. Sidereal time is then exactly 0h. 0m. 0s. for that meridian. Sidereal time is only of use for astronomers and astrologers. Since the beginning of sidereal day (0h.) can occur at any moment of the day or night according to the meridian and the time of the year it would be impractical to use this time system for ordinary civil purposes.

It is important that the astrologer understands sidereal time. With every chart he calculates he uses this measure of time. Looking at the column marked *Sidereal Time* in the yearly ephemeris, the figures in hours, minutes, and seconds for successive days can give a misleading impression of what they actually represent. For instance, around noon at Greenwich on the 22nd or 23rd March, at the vernal equinox, sidereal time is 0h. 0m. 0s. You can check this with *Raphael's Ephemerides* for any year. If you happen to have the 1965 ephemeris we can quote the actual figures. On page 6 the third column is headed *Sidereal Time*, and this time is given in hours, minutes, and seconds for noon *Greenwich Mean Time* for each day of March. On the 22nd sidereal time at noon was 23h. 59m. 07s.

On the 23rd sidereal time at noon was 00h. 03m. 03s. One will notice that throughout the year sidereal time at noon increases each day by roughly 4 minutes. Figures at noon on 23rd March 1965 increased on those for the 22nd by 3 minutes 56 seconds.

What exactly does this convey to the astrologer? It can tell him that this 3m. 56s. is the difference between the sidereal day and the mean solar day. It can tell him that this figure multiplied by the number of days in the year (365) will make a sum of approximately 24 hours in sidereal time, representing the one complete day which sidereal time gains on mean solar time in one year.

But the most important factor these increasing daily figures should convey to the astrologer is this. He knows that the Earth is for ever spinning on its axis. Thus, the difference of 3m. 56s. in sidereal time between noon on the 22nd March and noon on the 23rd, implies that in the 24 hours of *mean solar time* between both noons the Earth has rotated on its axis roughly 24h. 3m. 56s. of *sidereal time*. Because between the 22nd and 23rd sidereal time increased by 3m. 56s. this does not mean that the Greenwich meridian moved only that distance. The Greenwich meridian turned the full circle of 24 hours of sidereal time reckoning +3m. 56s. in the same period of *exactly* 24 hours of mean solar time reckoning.

We have said that sidereal time begins when the First Point of Aries crosses the observer's meridian. To this we must add that sidereal time is measured in the *opposite* direction to right ascension, *westwards* along the equator. Sidereal time is the *hour angle* of the First Point of Aries when expressed in time. Quoting the noon sidereal time figures again for 23rd March 1965 (00h. 03m. 03s.) this means that the hour angle measured westwards along the equator to the First Point of Aries from the Greenwich meridian was 00h. 03m. 03s. of sidereal time. On the previous day, noon, 22nd March, the hour angle measured

23h. 59m. 07s. westwards to the First Point of Aries from the Greenwich meridian.

Hour angle can be measured westwards to the First Point of Aries from the observer's meridian (along the equator); or westwards from the observer's meridian to a planet or any other point (Fig. 4).

One of the most important calculations in astrology is that of finding the *local sidereal time* for a given birthplace and a given time. *Greenwich Mean Time* was used throughout *Raphael's Ephemerides* until 1960, therefore the sidereal time given for noon on any day refers also to the *local* sidereal time at Greenwich.

You may understand this better if you realize that at *any moment* the sidereal time is different for two different meridians. When, for example, the sidereal time for the Greenwich meridian is 6h. 00m. 00s., this is also the sidereal time for all other places on the Greenwich meridian, irrespective of latitude. All other places on Earth, not exactly on the Greenwich meridian, which are intersected by other meridians from 0° to 180° westwards, or from 0° to 180° eastwards, would at that moment have different local sidereal times ranging over the full 24 sidereal hours. Sidereal time, when it is particular to a given meridian, is interpreted as *local* for that meridian.

Thus, when you are calculating the local sidereal time for a given time and place of birth you are determining the hour angle this particular meridian is westwards from the First Point of Aries. From this measurement of sidereal time which is *local* or particular to the birthplace meridian, you can find the culminating degree of the ecliptic (M.C.) and the corresponding Rising Degree (Ascendant) according to the latitude of the birthplace.

Solar Time, Mean Time

Apparent solar time is time reckoned by the true Sun, its apparent speed of motion round the Earth. This, in a sense, is *true* time, as registered by sundials. It has been called "God's time", because it is the natural measurement of the periods of day and night.

Apparent local solar time, for a given place and instant, is the interval since the Sun crossed the meridian (upper meridian if p.m., lower meridian if a.m.) at that place.

Mean time or *mean solar time* derives from the fictitious or mean Sun (p. 39). It is the time to which clocks are adjusted, because it provides a uniform measurement throughout each day and throughout the year. Hence, it is called *clock time*. Clock or mean solar time can also refer to both civil time and astronomical time. *Civil time* is clock time beginning at midnight (see *civil day*, p. 40). *Astronomical time* is clock time which began at noon (prior to 1925), but since then has begun at midnight (see *astronomical day*, p. 40).

On page 23 we read how 15° arc of terrestrial longitude is the distance travelled by the *mean* Sun in 1 hour of mean solar time. With this correspondence in time and longitude we can calculate what is the *local mean solar time* for some other meridian whose longitude is known, which is different from that of our own meridian. The difference in longitude we convert into time at the rate of 15° = 1 hour, or 1° = 4 minutes. We must remember the rule that as the Earth rotates from west to east, the more easterly meridian will be transited by the mean Sun first. Therefore, if a given meridian is *east* of one's own meridian, time for places on that meridian will be in *advance* of one's own local mean solar time. The reverse applies if a given meridian is *west* of one's own meridian; e.g. when it is noon on one's own meridian the local mean solar time for a place whose meridian is 20° *west* of one's own meridian will be 20° × 4 minutes = 80 minutes (1 hour 20 minutes), which subtracted from 12 noon = 10.40 a.m. For a place 20° *east* the local mean solar time will be 1.20 p.m.

This calculation is similar to *longitude equivalent in time*, one of the steps made by the astrologer in calculating the local sidereal time at birth. It refers to the difference in *time* between the place of birth and Greenwich, determined by their difference in longitude. Each 1° in longitude a place is *east* of Greenwich, 4 minutes must be *added*; or 4 minutes *subtracted* for each 1° *west*.

Greenwich Mean Time

In 1884 the meridian of Greenwich was chosen as the prime meridian from which all terrestrial longitudes should be measured. It was also recommended that a system of *Standard Times* should be introduced, differing by an integral number of hours from *Greenwich Mean Time*. Although these recommendations were not immediately adopted by all countries, practically all clocks throughout the world are now synchronized with Greenwich Mean Time (G.M.T.). Astronomically it is often called *Universal Time* (U.T.).

Greenwich Mean Time is, of course, *Greenwich local time* or *Greenwich civil time* for the meridian of Greenwich. It is also the standard time for the United Kingdom. But to get this perfectly clear it must be realized that, for example, 6.0 p.m. G.M.T. is the standard time the clocks throughout the United Kingdom are synchronized to, yet it is not the *local time* (or *local mean solar time*) for any other place in the U.K. except those on the Greenwich meridian. The local mean solar time for Liverpool, which is 3° longitude west of Greenwich, would be 3° × 4 minutes = 12 minutes behind G.M.T. Thus, when it is 6.0 p.m. at Greenwich, it is still only 5.48 p.m. at Liverpool in terms of the mean Sun. It is as well to understand this, though since 1884 all births in the U.K. would have been recorded in G.M.T. and not local mean solar time.

Equation of Time

Equation of time refers to the addition or subtraction to be made to convert apparent solar time (measurement of the true Sun) into mean solar time (measurement of the mean or fictitious Sun). The Earth moves round the Sun with variable speed, therefore the apparent speed of the Sun as viewed from the Earth varies at different times of the year. *Whitaker's Almanack* gives the value of the equation of time for 0h. of every day in the year. These values show that from around 25th December until mid-April, and between mid-June and end of August, the true Sun (sundial time) is ahead of the mean Sun (clock time). Whereas between mid-April and mid-June, and end of August and about 25th December, the mean Sun is ahead of the true Sun. The difference reaches as much as 16m. 20s. around 3rd November. At only four times in the year is the speed of the true Sun identical to that of the mean Sun: on or about 16th April, 14th June, 1st September, and 25th December.

Standard Time, Zone Time

An astrologer must always check to see whether a birth *outside the United Kingdom* is to a given *Standard Time*. As has been stated, most countries have now adopted Standard Time based on Greenwich Mean Time. For very large continents, such as the United States of America, or at sea, a system of *zones* is used to keep the time the same over particular ranges of longitude. These are called *time zones*. For instance, Eastern Standard Time (E.S.T.) is used throughout the time zone 67° 05′–82° 05′ west of Greenwich. E.S.T. is 5 hours slow of (behind) G.M.T. Pacific Standard Time (P.S.T.) is used throughout the time zone west of 112° 05′, and the clocks here register time 8 hours slow of G.M.T. When it is noon at Greenwich, it is 7 a.m. E.S.T. in New York, and 4 a.m. P.S.T. in

Seattle. Although these specified areas are called time zones, we speak of Standard Time on *land*, and *Zone Time* at *sea*.

Standard Time is equal to *local civil time* at adopted central meridians. These meridians are usually 15° apart, and central in the time zones. Between neighbouring zones, time changes abruptly. On opposite sides of the 180° meridian, time changes by a whole day. The variations in time between zones are in multiples of $\frac{1}{2}$ hour or 1 hour.

Summer Time

Summer Time was introduced into Great Britain in 1916, and has been used ever since. Many other countries have adopted this so-called *Daylight Saving Time*, for which the clocks are advanced 1 hour during several months between spring and autumn. During the War years in Great Britain *Double Summer Time* was used during summer, and for the rest of the year ordinary Summer Time applied.

Ephemeris Time

Ephemeris Time (E.T.) is the theoretical uniform time system employed by astronomers in gravitation theories of the Sun, Moon, and planets. It is used alongside Universal Time. Early in 1903 E.T. was the same as G.M.T., but gradually the two have diverged, and at the present time (1965) E.T. is 35 seconds in advance of G.M.T. Although *Raphael's Ephemerides*, which are the most popular reference books of this kind, have been calculated in E.T. since 1960, this 35 seconds discrepancy is going to cause scarcely any practical error for the astrologer who may treat the times given in the ephemeris as G.M.T. However, for correctness, 35 seconds should be *added* to any given G.M.T. to convert this to E.T.

STELLAR AND PLANETARY FACTORS

9 Stellar Systems and Galaxies

It is impossible for our human minds to comprehend the scale of the universe in terms of its stellar population and the colossal distances involved.

Galaxies are complex organizations of stars and inter-stellar matter (dust and gas). There are perhaps ten thousand million galaxies in the observable universe. Their size and structure vary considerably, and an average galaxy may have a population of a thousand million stars.

Inconceivable though the proportions may be, we can-not but recall the ancient Hermetic axiom, "as above, so below". The macrocosm (great world) is repeated in the microcosm (little world). From the basic structure of the atom with its elementary particles of proton, neutron, electron, to the infinite universe as a whole, there seems to be a mysterious repetition of patterns descending from the greatest to the minutest scale; cycles of birth, growth, and decay to which all matter is subject. All is energy, waxing or waning.

The astronomers speak of "aggregations of matter". The largest known aggregations are *clusters of galaxies*. Dense groups of stars are *star clusters*.

Nebulae is the name for cloud-like aggregations of stellar matter, usually referred to star-clusters so distant from our Earth that we see them as mere blots or wisps of "cloud". What in effect may be a pinprick of light to the naked eye could be a remote galaxy containing millions of stars!

Stars are self-luminous (as with our Sun) and are wholly

gaseous, unlike the planets, which are solid matter. There are stars known to be 3,000 times the size of our own Sun.

It is important that the beginner in astrology knows what is meant by *constellation* as distinct from *zodiacal sign* (p. 25). The sceptic of astrology invariably tries to debunk the subject by pointing out that, for instance, when the Sun is in Aries (sign) it is actually in Pisces, because of Precession. With regard to the actual constellation directly behind the Sun (as seen from the Earth) at the vernal equinox this is so. But the constellation Aries is *not* the sign Aries. The astrological or zodiacal signs refer to the twelve divisions of the zodiac which begin *always* at the vernal equinox with Aries. The twelve constellations of the same names refer to actual groups of stars. Constellations, generally, refer to arbitrary groups of stars conveniently chosen and named for the purpose of locating individual stars.

The Galaxy

The galaxy is the galactic system to which the Sun and its system of satellites (including, of course, our Earth), and most of the observable *individual* stars, belong. It is one of the largest known galaxies, containing perhaps 100 thousand million stars. It is now known to have a spiral structure, and is flattened in shape (looked at edge-on it bulges in the centre and tapers off at the extremities of its circumference).

It has been called the "Milky Way system" because of the encircling hazy band of light called the Milky Way that can be seen with the naked eye. This band is actually composed of stars, nebulae, and interstellar dust.

10 *The Solar System*

The Sun is the centre of its own system of satellites (planets), and as such it is the common centre of gravity for all bodies within this *Solar System*. It is the only self-luminous body in this system and is virtually the whole source of light and heat.

Besides the Sun and the nine larger planets, the Solar System contains several thousand minor planets or asteroids, satellites of planets such as the Earth's moon, comets, meteors, Zodiacal Light particles, and inter-planetary dust. The nine larger planets are dealt with in Chapter 11, and their orbits are shown in Fig. 1.

Asteroids or Minor Planets

Asteroids are the thousands of small bodies, most of which orbit the Sun between Mars and Jupiter. They are sometimes, though less commonly, called *minor planets* or *planetoids*.

Nearly all are too small to be seen with the naked eye. The largest is Ceres, its diameter a mere 480 miles. Ceres was the first asteroid discovered, on 1st January 1801. Most have a diameter of between 9 and 50 miles, some even smaller. It is possible that the total mass of all the several thousand asteroids is less than 1 per cent the mass of the Earth.

One theory of their origin is that they are the remains of a disintegrated planet which once orbited between Mars and Jupiter, particularly as the asteroids' mean orbit distance from the Sun closely fits Bode's empirical law of planetary distances (p. 57).

Comets

Comets are bodies orbiting the Sun, most of which are characterized by an extended luminous tail which is only visible when near the Sun. The tail always points away from the Sun. Most comets orbit in a direct motion, from west to east. Those with *elliptical* orbits return at regular intervals; but about three-fourths of all known comets have *parabolic* orbits (infinitely long ellipse), and these leave the Solar System for ever. A comet consists of stony particles or clouds of meteors. Its head or *coma* (which has a bright *nucleus*) may be from 30,000 to 150,000 miles across, and its tail from 5 million to 50 million miles in length.

Meteors, Meteorites

There are unknown millions of *meteors* scattered throughout the Solar System, all having elliptical orbits round the Sun. A meteor can be seen as a "shooting star", when it is momentarily made luminous by friction and consumed on entering the Earth's atmosphere at high speed—even though an average meteor is between a grain of sand and a large pea in size!

Meteors are usually divided into two classes: shower and sporadic. Shower meteors come from well-defined streams or swarms, such as the Leonids appearing each year in the middle of November when the Earth "passes through" the stragglers belonging to a particularly spectacular swarm which return *en masse* about every 33 years. Sporadic meteors appear from any direction at any moment.

A meteor that is not entirely consumed in the Earth's atmosphere, which strikes the Earth's surface, is called a *meteorite*.

The Zodiacal Light

Thirty per cent of the light of a moonless night sky reaches us from the *Zodiacal Light*, a zone of faint illumination that runs around the ecliptic. It consists of small rocky and reflecting particles, densest and brightest near the Sun.

Solar Apex

The Sun, carrying with it planets, asteroids, and the millions of meteor fragments, revolves round the *galactic centre*, in common with all the other millions of stars in our galaxy. At a speed of 12 miles per second its orbit takes it in the direction of the constellation Hercules, in right ascension 277° and declination 30° north, its position relative to the Earth at the end of December annually—when the Earth is then directly behind the Sun in its motion round the centre of the galaxy. The point to which the Sun's galactic motion is directed is called the *solar apex*.

11 *The Planets*

Five planets, excluding the Earth, were known to the ancients: Mercury, Venus, Mars, Jupiter and Saturn. It was not until 13th March 1781 that another planet was discovered. This was Uranus, which in many almanacs is still called by its first name, *Herschel*, after its discoverer, Sir William Herschel.

Astrologers refer to the three known planets beyond the orbit of Saturn (Uranus, Neptune, Pluto) as the *extra-Saturnian planets*.

Astronomers refer to Mercury and Venus as *interior planets*. They are the only planets whose orbits are inside the orbit of the Earth. They are more commonly termed *inferior planets*, in the sense that their orbits are smaller than the Earth's. The other planets outside the Earth's orbit are called *superior planets*.

Between the orbits of Mars and Jupiter are the hundreds of *minor planets* or asteroids (see Chapter 10). The four planets inside this asteroid belt (Mercury, Venus, Earth, Mars) are sometimes spoken of as the *inner planets*. The five planets outside this belt (Jupiter, Saturn, Uranus, Neptune, Pluto) are called the *outer planets*. Four of these, excluding the fifth, Pluto, are also known as *major planets*, because they are all large, of low density, and surrounded by atmospheres. Whereas the inner planets are relatively much smaller and of greater density. Mercury, Venus, Earth, Mars and Pluto are also grouped together as the *terrestrial planets* in terms of their physical properties being somewhat similar, each appearing to be of solid matter like the Earth.

In Fig. 1 (top) the planets are described in their order of nearness to the Sun, though the diagram is of necessity greatly out of proportion. An interesting feature of their relative distances from the Sun was first recognized by an astronomer called Bode. This was in 1772, before the discovery of Uranus, and yet what has come to be known as "Bode's Law" does seem to apply closely to the asteroids (first discovered in 1801) and Uranus. But his calculations applied to Neptune would seem to predict Pluto's orbit much better. Bode's Law is as follows: beginning with Mercury, write down the figures 0, 3, 6, 12, 24 (that is, doubling up) . . . add 4 to each sum, and divide by 10. In the table below, the numbers thus obtained are given in the third column. In the fourth column are the mean distances of the planets from the Sun, in terms of the Earth's distance (called the *Astronomical Unit*).

Planet	*Series*			*Bode's Law*	*Ast. Units*
Mercury	0	add 4,	÷ 10	0·4	0·39
Venus	3	add 4,	÷ 10	0·7	0·72
Earth	6	add 4,	÷ 10	1·0	1·00
Mars	12	add 4,	÷ 10	1·6	1·52
Asteroids	24	add 4,	÷ 10	2·8	2·65
Jupiter	48	add 4,	÷ 10	5·2	5·20
Saturn	96	add 4,	÷ 10	10·0	9·54
Uranus	192	add 4,	÷ 10	19·6	19·19
Neptune	384	add 4,	÷ 10	38·8	30·07
Pluto	768	add 4,	÷ 10	77·2	39·52

Before listing a few interesting facts and figures concerning each planet (for convenience we are classifying the Sun and Moon as planets) the following definitions may be helpful in giving the reader a clearer understanding of differences in the constituents of the planets:

Sidereal period: the planet's sidereal year.

Synodic period: period between two consecutive conjunctions with the Sun, viewed from the Earth.

Mean orbital speed: mean speed along the ecliptic.

Orbital eccentricity: expresses the elliptical shape of the orbit. For instance, Saturn's is 0·056; the Earth's is 0·017. Saturn's orbit is, therefore, nearly $3\frac{1}{2}$ times as elliptical as the Earth's, indicating that the Earth's path around the Sun is less eccentric, nearer to being a circle, than is the case with Saturn's orbit. A circle is expressed as zero eccentricity, 0·00. All ellipses have eccentricities between zero and 1·00.

Orbital inclination: describes *how the orbit is tilted* relative to the plane of the ecliptic. In other words, the maximum *latitude* the planet can reach. This value for the Earth is zero, since the ecliptic is the Earth's orbit.

Earth diameters: the planet's diameter in terms of the Earth's, which is expressed as 1·00. Mercury's diameter is 0·39 = 39 per cent that of the Earth's.

Earth volumes: equivalent volume to that of the Earth's, which is expressed as 1·00. Volume is the number of cubic miles of space the planet contains, whatever matter or gas the body consists of.

Earth masses: equivalent mass to that of the Earth's, which is expressed as 1·00. Mass is the quantity of matter within the planet's structure.

Earth densities: equivalent density to that of the Earth's, which is expressed as 1·00. Density is defined as the number of times its mass exceeds that of a sphere of pure water having the same dimensions.

Surface gravity (Earth's): equivalent surface gravity to that of the Earth's, which is expressed as 1·00. Surface gravity specifies the velocity of escape from a planet. A combination of temperature and surface gravity determines whether a planet can retain an atmosphere or not.

Length of day: planet's period of rotation on its axis.

The Sun

Earth diameters:	109
Earth volumes:	1,300,000
Earth masses:	333,420
Earth densities:	0·26
Surface gravity (Earth's):	28·0

Although the Sun's volume is over $1\frac{1}{4}$ million times greater than that of the Earth, its mass is just over 300,000 times the Earth's mass, therefore its density can only be about $\frac{1}{4}$ of the Earth's density. The Sun, being a star, is not a sphere of solid matter. Its surface is never still, but is a ceaselessly boiling inferno of swirling, spurting gases.

The Sun rotates on its axis in the same direction as the planets revolve, eastwards. This rotation of the Sun was discovered through the use of telescopes, when man was able to plot the movements of spots on the Sun's surface going around the disc, disappearing, and appearing again periodically. As the Sun is composed of gases its whole sphere does not rotate in the same period. It is estimated that at the Sun's equator the mean rotation period is about 24·65 days. With each degree of latitude, north or south, towards the poles, the rotation period increases, until at the poles the period is as much as 34 days. The Sun's axis is inclined to the ecliptic at an angle of 82° 49·5′.

The study of sunspots can be of tremendous interest to the astrologer. There are records of spots being observed by the ancients with the naked eye, but it was not until around 1600 that Galileo first saw them through his telescope. No serious study of their growth and possible effects on terrestrial life was undertaken until about 1700. Sunspots follow a cycle of roughly 11·13 years. This cycle has maximum and minimum phases which occur alternately. At maxima there can be as many as 200 spots at one time, most lasting a few days, some for weeks. A group of spots may extend over more than 200,000 square miles of the Sun's surface and emit flames (solar flares) thousands of miles high. It is recognized that sunspots possess powerful magnetic fields, and that these solar eruptions very definitely affect the ionosphere surrounding the Earth, with resultant affects on the Earth's magnetism, often upsetting radio communications. Extremely interesting attempts have been made to connect cycles of human and wild-life activity with the sunspot cycles.

Mercury

Mean distance from Sun:	36 million miles
Sidereal period:	87·97 days (0·2408 tropical years)
Synodic period:	115·88 days
Mean orbital speed:	30 miles per second
Orbital eccentricity:	0·206
Orbital inclination:	7° 0′
Earth diameters:	0·39 (39%)
Earth volumes:	0·06 (6%)
Earth masses:	0·04 (4%)
Earth densities:	0·69 (69%)
Surface gravity (Earth's):	0·27 (27%)
Max. surface temperature:	770°F.
Length of day:	88 Earth days
Inclination of equator to orbit:	uncertain
Known moons:	none

It is the planet nearest to the Sun, as might be deduced from its temperature which can reach 770°F.—hot enough to melt lead! Yet as Mercury always turns the same face to the Sun one side is in perpetual shadow, and the temperature on this side is thought to be as low as − 273°C., even colder than the remote Pluto.

Thus, Mercury can claim a number of unique features. It is at the same time the hottest and the coldest among the planets; it is smallest in size and mass; has the shortest sidereal period; greatest linear speed; receives most light from the Sun; and has the most eccentric, and less nearly circular, orbit, except for Pluto's.

Mercury appeared in Babylonian records as far back as the 4th century B.C. It can never be more than 28° from the Sun, as seen from the Earth. This occurs at Mercury's aphelion at elongation; but at its perihelion at elongation its maximum angle from the Sun may be as small as 18°.

Venus

Mean distance from Sun:	67·2 million miles
Sidereal period:	224·7 days (0·6152 tropical years)
Synodic period:	583·9 days
Mean orbital speed:	22 miles per second
Orbital eccentricity:	0·0068
Orbital inclination:	3° 24'
Earth diameters:	0·97 (97%)
Earth volumes:	0·92 (92%)
Earth masses:	0·82 (82%)
Earth densities:	0·89 (89%)
Surface gravity (Earth's):	0·86 (86%)
Max. surface temperature:	140°F.
Length of day:	247 Earth days
Inclination of equator to orbit:	uncertain
Known moons:	none

Venus is the brightest object in the sky after the Sun and Moon. It has the most nearly circular orbit (0·006 eccentricity) of any planet as it revolves round the Sun. From superior conjunction (far side of Sun from Earth) to greatest elongation it takes 220 days, but the interval from greatest elongation to inferior conjunction (near side of Sun from Earth) is only 72 days. The farthest it can be from the Sun, as seen from the Earth, is 48°, at elongation. Five synodic revolutions (successive conjunctions with the Sun) take 8 years, to within a day.

Its dense, opaque atmosphere prevents the planet's

surface being seen, hence the uncertainty regarding its
rotation period. A recent (1964) theory puts this period as
247 days. If this is so, it would be unique for a planet's
rotation period (247 days) to be longer than its revolution
period (225 days). It is also suggested that, like Uranus,
the rotation of Venus is retrograde—in the opposite
direction to its eastward journey round the Sun.

Earth

Mean distance from Sun:	92·9 million miles
Sidereal period:	365·26 days
Synodic period:	0
Mean orbital speed:	18½ miles per second
Orbital eccentricity:	0·017
Orbital inclination:	0
Max. surface temperature:	140°F.
Inclination of equator to orbit:	23° 27′

The third planet from the Sun, and the nearest to have
a satellite. Only Venus and Neptune have less eccentric
orbits. Its orbit, of course, is not inclined to the ecliptic—
its orbit *defines* the ecliptic. The axis of rotation is
inclined to the ecliptic by 66° 31′ 01″, which produces the
obliquity of the ecliptic relative to the equator. With the
possible exception of Pluto, the Earth is the densest known
body in the Solar System.

The Moon

Mean distance from Earth:	238,857 miles
Mean sidereal period:	27d. 7h. 43m. 11·47s. (27·32 days)
Mean synodic period:	29d. 12h. 44m. 2·78s. (29·53 days)
Mean orbital speed:	0·63 miles per second

Orbital eccentricity:	$\frac{1}{18}$th
Orbital inclination:	5° 18′ (mean 5° 8′)
Earth diameters:	0·273 (27%)
Earth volumes:	0·02 (2%)
Earth masses:	$\frac{1}{81}$
Earth densities:	0·61 (61%)
Surface gravity (Earth's):	0·16 (16%)
Max. surface temperature (at lunar mid-day):	373°
Min. surface temperature (at lunar midnight):	193°
Inclination of equator to orbit:	6° 5′

Next to the Sun the Moon is the brightest object in the sky for we on Earth, but it reflects only 7 per cent of the sunlight that falls upon it. It is a relatively large satellite, being ¼ the size of the Earth, more than ⅔ as large as Mercury.

Mars

Mean distance from Sun:	141·5 million miles
Sidereal period:	687 days (1·88 tropical years)
Synodic period:	779·9 days
Mean orbital speed:	15 miles per second
Orbital eccentricity:	0·093
Orbital inclination:	1° 51′
Earth diameters:	0·53 (53%)
Earth volumes:	0·15 (15%)
Earth masses:	0·11 (11%)
Earth densities:	0·70 (70%)
Surface gravity (Earth's):	0·37 (37%)
Max. surface temperature:	86°F.
Length of day:	24h. 37m. 23s.
Inclination of equator to orbit:	25° 12′
Known moons:	two

The oppositions of Sun–Mars, when Mars is at perihelion, occur in the latter part of August every 15–17 years. Distance from Earth is then only about 36 million miles. When Mars is in opposition at aphelion the distance can be as much as 61 million miles. When Sun–Mars are in conjunction the distance is about 234 million miles. (Conjunction and opposition are illustrated in Fig. 20.)

In its *synodic year* (two successive conjunctions with Sun as seen from Earth) of 780 days, Mars is direct for 710 days, and retrograde for 70 days. This synodic period is the longest for any of the planets, due to the small difference between the sidereal periods of Earth and Mars.

Mars has been called the "red planet" due to its reflecting red light better than blue light, hence it looks red. The most controversial features associated with Mars are the so-called "canals". They were originally called *canali* by the Italian observer Schiaparelli, and a mistranslation led to the belief that intelligent beings had constructed these "canals", possibly for irrigation.

Jupiter

Mean distance from Sun:	483·3 million miles
Sidereal period:	11·86 tropical years
Synodic period:	399 days (1·092 tropical years)
Mean orbital speed:	8 miles per second
Orbital eccentricity:	0·048
Orbital inclination:	1° 18′
Earth diameters:	10·97
Earth volumes:	1,318
Earth masses:	318·3
Earth densities:	0·24
Surface gravity (Earth's):	2·64
Max. surface temperature:	−216°F.
Length of day:	9h. 55m.
Inclination of equator to orbit:	3° 7′
Known moons:	12

Jupiter has a mass of $2\frac{1}{2}$ times that of all the planets put together. Its volume, too, is greater than the total of all the planets. Yet its density is less than $\frac{1}{4}$ that of the Earth. Its day is shorter than any of the planets, hence its rotation is the most rapid. The unique feature of the planet's surface is the Great Red Spot, which has been seen in varied forms for over 100 years, and is thought to be the result of a volcanic eruption on the planet. An interesting factor, too, is its system of twelve moons. The four brightest were discovered by Galileo. A remarkable thing about the four outermost moons is that they revolve in a *retrograde* motion. Only one of Saturn's moons also has this motion. It is thought they are asteroids "captured" by Jupiter's great gravitational pull.

Saturn

Mean distance from Sun:	886·1 million miles
Sidereal period:	29·46 tropical years
Synodic period:	378 days (1·035 tropical years)
Mean orbital speed:	$6\frac{1}{2}$ miles per second
Orbital eccentricity:	0·056
Orbital inclination:	2° 29′
Earth diameters:	9·03
Earth volumes:	736
Earth masses:	95·3
Earth densities:	0·13
Surface gravity (Earth's):	1·17
Max. surface temperature:	−243°F.
Length of day:	10h. 38m.
Inclination of equator to orbit:	26° 45′
Known moons:	9

Saturn has only 13 per cent of the Earth's density. So low, in fact, that it is the only planet that would float in water! Its outermost satellite orbits Saturn in a *retrograde* motion. Saturn's unique feature is the system of rings which are exactly in the plane of the equator. It is thought they are composed of rocky fragments, and probably the remains of an exploded satellite. The outer ring has a diameter of 275,000 km.

Uranus

Mean distance from Sun:	1,783 million miles
Sidereal period:	84·02 tropical years
Synodic period:	370 days (1·012 tropical years)
Mean orbital speed:	4 miles per second
Orbital eccentricity:	0·047
Orbital inclination:	0° 46′
Earth diameters:	4·00
Earth volumes:	64
Earth masses:	14·7
Earth densities:	0·23
Surface gravity (Earth's):	0·92
Max. surface temperature:	300°F.?
Length of day:	10·7h.
Inclination of equator to orbit:	98°
Known moons:	5

With Uranus its axis is nearly in the plane of the orbit: the inclination of the equator to plane of its orbit is 98°. Thus, its axial rotation is retrograde. That is, the axial rotation of the other planets is eastwards, the same direct motion of their orbit round the Sun. Because of this peculiarity there would be no season on Uranus as we know them. Uranus alternately presents its poles and equator to the Sun, so that each pole would experience a 42-year "summer", and a 42-year "winter" in its sidereal period of 84 years. Discovered in 1781.

Neptune

Mean distance from Sun:	2,797 million miles
Sidereal period:	164·79 tropical years

Synodic period:	367·4 days (1·006 tropical years)
Mean orbital speed:	3⅓ miles per second
Orbital eccentricity:	0·0086
Orbital inclination:	1° 47'
Earth diameters:	3·90
Earth volumes:	39
Earth masses:	17·3
Earth densities:	0·29
Surface gravity (Earth's):	1·44
Max. surface temperature:	330°F.?
Length of day:	15·8h.
Inclination of equator to orbit:	29°
Known moons:	2

Is the first planet (from the Sun) to deviate to any large extent from the prediction of Bode's Law (p. 57). Its orbit is almost circular. Discovered in 1846.

Pluto

Mean distance from Sun:	3,670 million miles
Sidereal period:	248·4 tropical years
Synodic period:	366·7 days (1·004 tropical years)
Mean orbital speed:	3 miles per second
Orbital eccentricity:	0·249
Orbital inclination:	17° 19'
Earth diameters:	0·46
Earth volumes:	0·10
Earth masses:	same?
Earth densities:	?
Surface gravity (Earth's):	?
Max. surface temperature:	348°F.?
Length of day:	6d. 9h. 16m. 54s.
Inclination of equator to orbit:	?
Known moons:	none

Because of the large eccentricity, part of Pluto's orbit lies inside that of Neptune. But the large inclination of its orbit to the ecliptic (17° 19′) makes a collision between these planets practically impossible. It has been suggested that Pluto is merely an ex-satellite of Neptune, which struck an independent orbit.

However, recent observations (1965) now suggest that Pluto may be much larger than previously thought—considerably larger than the Earth, and that it is an independent planet in its own right. Its mass is thought to be enough to produce the perturbing effects on Uranus and Neptune which led to its discovery in 1930. Its apparent small appearance is believed to be an effect of specular reflection; its disc is brightest at its centre, and darkens towards the limb, and its magnitude varies regularly over a period of a few days. At such an immense distance its small size has therefore been an illusion.

ORBITAL ELEMENTS

12 *The Earth's Motions*

Axial Rotation

The two principal motions of the Earth are:
1. Axial rotation.
2. Orbital revolution.

Rotation means the spinning of a body on its axis. It is because of this motion of the Earth that the whole sky appears to turn about the Earth, and we experience the successive periods of day and night. We speak of the Earth's *diurnal rotation*, which is performed in the period called a sidereal day. This motion is from west to east. That is, if you stand facing south the Earth beneath your feet turns eastwards, anti-clockwise.

Orbital Revolution

The orbital motion of the Earth about its centre of attraction (the Sun) is called its *orbital revolution*. The Earth, like all the planets, moves eastwards around the Sun. This so-called *direct* motion follows the eastward rotation of the Sun on its own axis. The Earth, however, does not move around the Sun in an exact circle: its orbit is very nearly an ellipse.

Kepler's Laws of Planetary Motion

Fig. 6 illustrates the orbital motion of the Earth. This is based on Kepler's Laws of Planetary Motion. It was in 1609 when Johannes Kepler gave to the world the first two of his three laws of planetary motion. The third was announced in 1618. These three laws began a new era in the history of astronomy. They are as follows:

1. *The orbit of a planet is an ellipse, with the Sun at one focus.* This first law describes the *shapes* of the orbits, and defines the path which the planet always follows. All the elliptic orbits of the planets have one focus in common, and that lies at the centre of the Sun. Fig. 6 shows the Sun

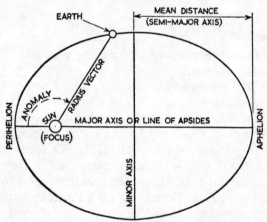

Fig. 6. The Earth's Elliptical Orbit

situated at one of the foci. Wherever the planet is in its orbit its distance from the Sun's centre is called the *radius vector*. The radius vector applies to a straight line drawn at any moment from a moving body to the centre round which it moves. Thus, in a circular orbit the radius vector is constant, but in an elliptic orbit it varies in length with the position of the moving body. As we read on page 58, *orbital eccentricity* expresses the elliptical shape of the orbit.

2. *A line (radius vector) drawn from a planet to the Sun sweeps over equal areas in equal times.* This second law is known as Kepler's law of areas. It describes the *varying velocity* or *speed* of motion of the planet at different points in its orbit. This is illustrated in Fig. 7. *Perihelion* (Gk: *peri*, near; *helios*, Sun) is the point on the orbit where

the radius vector is shortest. In the Earth's case this occurs annually around the beginning of January, which means the Earth is then at its closest to the Sun. *Aphelion* (Gk: *ap*, away from; *helios*, Sun) is where the radius vector is longest, and for the Earth this occurs annually around the

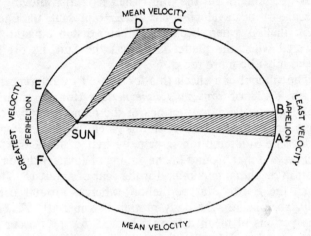

Fig. 7. Kepler's Law of Equal Areas

beginning of July, and means the Earth is then at its farthest from the Sun. The longest axis of the ellipse, the line joining perihelion and aphelion, is the *line of apsides*. The *minor axis* or shortest diameter of the ellipse crosses exactly the mid-point between perihelion and aphelion (Fig. 6). This midpoint measures the *mean distance* of the Earth from the Sun. The *anomaly* is the angle between the Earth's radius vector and the longest axis, measured from perihelion in the direction of motion, and gives the position of the Earth in its orbit.

Fig. 7 shows three shaded portions of the ellipse, representing three *equal* areas. It is apparent that where the radius vector is longest, the distance travelled by the Earth along its orbit is shortest, and *vice versa*. This is necessary to produce the equal areas. The arc *AB* represents the Earth at aphelion, when, because it is at greatest distance from the Sun (its centre of attraction), it is moving at its slowest speed in its orbit, and therefore covers least distance. The arc *CD* represents the Earth at mean distance from the Sun, when it is also moving at average or mean velocity. The arc *EF* represents the Earth at perihelion, when, because a greater distance has to be covered when the planet is nearest the Sun, its orbital speed must be more rapid.

The student can check this for himself by reference to pages 26–28 of *Raphael's Ephemerides* (for years since 1905) where the daily motions of the planets throughout the year are given. This motion is given in celestial longitude. The Sun's motion is actually that of the Earth. It can be seen that around the beginning of January the daily motion reaches a maximum for the year of about 1° 1′ 11″ when the Earth is at perihelion; whereas around early July, at aphelion, the daily motion is a mere 0° 57′ 11″. The periods of mean motion (about 0° 59′ 11″) occur at the beginning of April and October respectively.

3. *The square of the time that each planet takes to complete its orbit varies as the cube of the semi-major axis of that orbit.* This third law establishes a relation between the mean distance of a planet from the Sun and the periodic time (sidereal year), or, put another way, the *size* of the orbit to the time of revolution. Given its mean distance from the Sun one can find out the exact period of a planet's orbit. It also states a relation between certain elements of *two or more planets*. For example, using the Earth's mean distance from the Sun as a basic unit, and given the mean distance of Uranus, we can calculate the period of Uranus' orbit round the Sun. The mean distance of Uranus is about nineteen times that of the Earth. We find the square root of 19, which is 4·358 (say 4·4). Multiply 19 by its square root, 19 × 4·4 = 83·6 years. On

page 66 we read that Uranus takes about 84 years to orbit the Sun.

Gravitation

Kepler's three laws of planetary motion are independent of one another. It was left for Sir Isaac Newton sixty years later to show that all three could be deduced from one single law: the law of *gravitation*. Newton's law states:

Every particle of matter in the universe attracts every other particle with a force varying directly as the product of their masses and inversely as the square of the distance between them.

Gravitation derives from the Latin, *gravitas*: weight. Gravity acts upon objects to produce a force. This force is experienced as weight, or as pressure. The central point of Newton's theory was the importance of *mass*, which is the quantity of matter in a body. The cause of variable motion is because matter has mass. Every particle of matter attracts every other particle. The Earth attracts the Sun in the same way as the Sun attracts the Earth. But the Sun, having the greater mass, has the greater attractive force. If the Sun had no mass the planets would not move in ellipses. The planets move only because they are *acted on by some force*, otherwise they would be motionless, or move in a straight line.

Kepler's Second Law enables astronomers to determine the *direction* of this force. By the application of his First Law the force *at different parts* of the same orbit of a planet can be *compared*. His Third Law makes it possible to *compare the forces on different planets*.

The terrestrial globe of our Earth attracts everything to itself, like a magnet. All over its surface, weight is directed toward its centre, following the law of gravity which is an effect produced by attraction to a centre. Weight, as we know it, is the tendency of bodies to fall towards the Earth.

Yet the Earth, not being an exact circle but flattened at the poles and bulging at the equator, does not produce the same gravitational attraction over the whole of its surface. The force of gravity *increases* as one moves from the equator toward the poles. At the equator the Earth turns fastest on its circumference, with *decreasing* speed as one approaches the poles.

Planetary Perturbations

We can see gravity as a controlling force, keeping planets and satellites in their orbits with regard to a central body, yet it is also a disturbing influence. If it were not for the gravitational attraction of the other planets, the Earth would constantly describe the same orbit round the Sun. We call the continual disturbing effects of the other planets *perturbations* of the Earth's orbit.

Precession of the Equinoxes

Another important motion of the Earth is that known as the *Precession of the Equinoxes*. This is caused by the disturbing attractions of the Sun and Moon on the protuberant matter at the Earth's equator. Precession is a slow swinging motion of the Earth's axis of rotation. As we know, the Earth's axis is tilted, and the angle so produced between the plane of the equator and the plane of the ecliptic is called the *obliquity of the ecliptic*. The northern extremity of the axis points towards a star situated in the constellation known as Ursa Minor or the Little Bear. This star is called Polaris or the Pole Star, and it is but a fraction off true north. Thus, as the Earth seems to revolve round Polaris this star appears to be fixed.

But Polaris was not always so near to the Earth's north pole. Due to Precession, the Earth rotates obliquely upon itself, seen as a gradual retrograde motion of the Equinoctial Points along the ecliptic against the background of stars. The Equinoctial Points *advance* each year on the *real* revolution of the Earth round the Sun by a mean velocity of 50″ of arc, from east to west, which appears as the *retrograde* motion against the background of constellations.

About 2,000 years ago, when this displacement of the Earth's axis was first discovered by the great observational astronomer, Hipparchus, the vernal equinox "fell in" the constellation Aries. Hence, the First Point of Aries. But due to Precession the First Point of Aries is, at the present time, in the constellation Pisces. The complete retrograde cycle of the Equinoctial Points takes about 25,800 years (called a *Great Year*). In effect it is the pole of the Earth's equator which describes a circle, with a radius of $23\frac{1}{2}°$, round the *pole of the ecliptic*. In 3000 B.C. Draconis was the Pole Star and the vernal equinox coincided with the constellation Taurus. In A.D. 7500 Cephei will be near the celestial pole, and in A.D. 14000, Vega. Polaris will again be Pole Star around A.D. 28000.

Nutation

The combined luni–solar forces which cause Precession do not act quite uniformly. The backward motion of the Moon's nodes along the ecliptic (p. 81), called *regression of the nodes*, completes a full cycle in about 19 years. For half this period the plane of the Moon's orbit is inclined to the ecliptic in the *same way* as the plane of the Earth's equator is, and its precessional effect upon the protuberance at the Earth's equator is then small. But when, during the other half, the plane of the orbit is considerably inclined from the plane of the Earth's equator, the Moon's effect upon Precession is great. This causes the pole of the Earth's equator to perform a waved or oscillatory, and not regularly circular, path around the pole of the ecliptic in the secular inequality of Precession of the Equinoxes.

Earth's Motions Given in Ephemeris

The main motions of the Earth can be traced from data given in *Raphael's Ephemerides*. The length of consecutive sidereal days (i.e. period of axial rotation) can be determined by figures given for noon at Greenwich. For instance, on 1st January 1965, sidereal time at noon for Greenwich was 18h. 43m. 42s. At noon on the 2nd January, 24 mean solar hours later, sidereal time was 18h. 47m. 39s. This is an increase of 3m. 57s. on the previous noon. This tells us that the length of the mean solar day (clock time) from noon 1st January to noon 2nd January equals 24h. 3m. 57s. of *sidereal time*. But the sidereal day beginning noon 1st January completed its full 24 *sidereal* hours approximately 3m. 57s. (clock time) *before* noon on the 2nd January.

The position of the Earth in its orbit of the Sun, relative to the First Point of Aries, can be deduced from the 4th column on the left-hand page for each month, under "⊙ Long." On the 1st January 1965 the longitude shows ♑ 10° 52′ 37″. This equals 280° 52′ 37″ in right ascension, which is the distance the Earth has travelled *eastwards* along the ecliptic since the vernal equinox (20th March 1964).

Since the plane of the Earth's path round the Sun defines the ecliptic the Earth cannot have latitude. But the angle at which the Earth's equator is tilted relative to the Sun, which varies with the Earth's changing direction round the Sun, is shown under the heading "⊙ Dec." in column 5 in the ephemeris. We also know this, of course, as the Sun's angle of declination to the Earth's equator, as viewed from the Earth.

The relative *speed* with which the Earth orbits the Sun can be found on pages 26–28 in *Raphael's Ephemerides* (since 1905), under the heading of the Sun's daily motion (see p. 74). The *rate* at which the Earth's equator changes its angle of tilt relative to the Sun can also be found by comparing successive days' motion of the Sun in declination. At the equinoxes this rate can be as much as 25' per day, but as each solstice approaches so the rate decreases, until around 21st–22nd June, or 21st–22nd December, the declination remains the same, it "stands still".

13 Moon's Motion and Phases

The Moon's motion is extremely complex due to the interplay of the Sun, Earth, and Moon, and their relative positions are *never* exactly repeated. Indeed, there are about 150 principal periodic motions along the ecliptic, and just as many perpendicular to it—plus about 500 smaller perturbations!

However, the astrologer need only be interested in the few principal perturbations we shall mention, which are primarily caused by the Sun's gravitational attraction on the Earth–Moon system.

Common Centre of Gravity

Perhaps first we should realize that the Moon does *not* revolve round the *centre* of the Earth. Both bodies revolve about a common centre of gravity. As the mass of the Earth is eighty-one times the mass of the Moon, their centre of gravity is eighty times nearer the centre of the Earth than it is to the centre of the Moon. This point round which they both revolve is, in fact, about 1,000 miles *below* the Earth's surface—roughly 3,000 miles from the Earth's centre. It is this centre of gravity of the Earth–Moon system, and not the centre of the Earth, that describes an elliptical orbit round the Sun.

In Fig. 8 this motion is described. The sphere of the Earth is shown in its relative position to the Earth–Moon common centre of gravity at each of the quarters of the lunation. It will be seen that this centre remains constantly at the same distance beneath the Earth's surface. At Full Moon the Earth is on the sunward side of the centre of gravity; at New Moon on the opposite side. Therefore, distinct from its annual ecliptical orbit of the Sun, the Earth, due to sharing this common centre of gravity with the Moon, also travels round a little circle of 3,000 miles radius in the period of a lunation, being farther from the

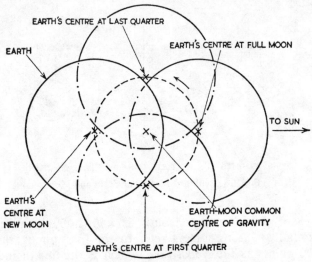

Fig. 8. Earth–Moon Common Centre of Gravity

Sun by this amount at New Moon and nearer at Full Moon. The effect of this is to make the Sun appear $6\frac{1}{2}''$ of arc in front of its mean place at the Moon's First Quarter, and $6\frac{1}{2}''$ behind it at the Last Quarter.

Moon's Nodes and Declination

Fig. 9 illustrates the combined movement of the Earth–Moon system. By its gravitational hold on the Moon the Earth appears to carry its satellite with it along the ecliptic. But, unlike the Earth, the Moon's orbit is not in the ecliptic, but inclines at an average angle of 5° 8′ with it. The two points in the Moon's orbit where it cuts the

plane of the ecliptic are called the *nodes*. The *ascending
node* (☊), called by the ancients the Dragon's Head, and
generally by astrologers the *north node*, is where the Moon
crosses the ecliptic from south to north. The *descending
node*, or *south node* (Dragon's Tail), where the ecliptic is

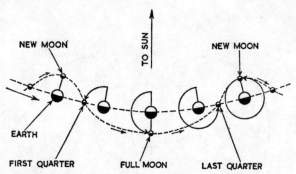

Fig. 9.   Combined Motion of Earth and Moon

crossed from north to south, is known by its reverse
symbol ☋. Astronomers refer to the line joining these
two points as the *Nodal Line*, which is the line of intersection of the plane of the Earth's path and that of the
Moon's path.

Fig. 10 describes the derivation of the Moon's nodes,
and also explains the angle known as *declination* in the
case of the Moon. On page 77 we read of the *regression
of the Moon's nodes*, which is a retrograde motion along
the ecliptic of about 19° in a year. This can be checked
from *Raphael's Ephemerides* for any year. For example, on
1st January 1965 Moon's north node cut the ecliptic at
♊ 21° 57'; on 31st December 1965 at ♊ 2° 41' = a retrograde distance of 19° 16' of arc. The line of nodes (Nodal
Line), with the plane of the Moon's orbit, performs a
complete retrograde revolution in 6793·4 days (about 18·6
years).

The angle of *inclination of the Moon's orbit* to the Earth's equator (declination) is subject to periodic variations due to this regression of the nodes, and because the Moon's orbit is not in the same plane as the Earth's equator. If

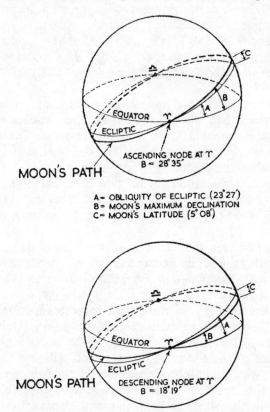

A = OBLIQUITY OF ECLIPTIC (23° 27′)
B = MOON'S MAXIMUM DECLINATION
C = MOON'S LATITUDE (5° 08′)

Fig. 10.   Moon's Nodes and Inclination of Orbit

you look through several ephemerides for consecutive years you will notice that the maximum declination attained by the Moon varies from month to month, year to year, reaching a maximum of perhaps 28° 35′ and a minimum of about 18° 19′ in a period of 19 years. When the Moon's orbit coincides with the First Point of Aries the ascending node will be at ♈ (Fig. 10), though at the vernal equinox the node's position may not be exactly ♈ 0°. The Moon then attains its *greatest* inclination (declination) to the Earth's equator during the 19 years' cycle, about 28° 35′. This figure is arrived at by adding the angle of the equator to the ecliptic (23° 27′), to the average angle of the Moon's orbit to the ecliptic (5° 8′). 23° 27′ + 5° 8′ = 28° 35′.

You can check this yourself if you have an ephemeris for a year when the Moon's north node reaches ♈ 0°. This last occurred in August 1950, and on 19th September maximum declination reached 28° 44′. About 9½ years later, when the ascending node intersects the autumnal equinox (♎), the Moon's inclination to the Earth's equator reaches its *lowest* in the 19 years' cycle. This last occurred in 1959, when the Moon's declination at maximum was as low as 18° 10′.

It can be seen that the maximum angle of declination attained by the Moon during a lunation depends upon the position of the nodes in the Moon's orbit to the First Point of Aries.

Raphael's Ephemerides can help us to see this nodal motion of the Moon more clearly. The 7th column on the left-hand page for any month contains the noon (G.M.T.) positions of the Moon in *latitude*. You will find that on the day when the Moon crosses the ecliptic from south to north, the longitude of the Moon at noon (6th column) will show the Moon to be in the same sign as that given at the top of the right-hand page for its north node. The Moon will be in the opposite sign (to the north node) when its latitude is shown to change from north to south. You might ask, if the Moon is only at its nodal points twice in a month, why are the node's positions given for alternate days? This is because the Earth's axial rotation brings the nodal point of intersection of the plane of the Moon's orbit with that of the ecliptic on to the upper meridian of a given place once every 24 hours, but in that time the nodal point will have retrograded an average of 3' of longitude.

See also definition of *nodical month*, page 36.

Perigee and Apogee

The Moon's orbit of the Earth observes the same law of equal areas as the Earth in its orbit of the Sun. The Moon's distance from the Earth, and speed in its orbit, varies. It is nearest to the Earth at *perigee*, farthest at *apogee*. At perigee the Moon's rate of motion in its orbit is fastest, at apogee slowest. The student will be interested to check this phenomenon with *Raphael's Ephemerides*. Since 1899 the perigee and apogee times of the Moon have been listed annually. Mostly this data has appeared on page 29. The daily motions of the Moon are given on pages 26–28. As an example, the Moon was in apogee 2nd January 1965, and the distance it covered in longitude (p. 26) was 11° 47'. Just after midnight 16th–17th January the Moon was in perigee, and for the 24 hours from noon 16th to noon 17th we find the Moon's motion in longitude was as much as 15° 14'.

The Moon's average maximum distance (apogee) from the Earth is about 252,000 miles; at perigee 225,000 miles; mean distance around 239,000 miles.

Line of Apsides

The line connecting the two points of the Moon's perigee and apogee, the longest axis of the ellipse traced by the Moon's orbit, is the *line of apsides*, or *apse line*. We speak of the *progression of the line of apsides*, an effect causing the Moon's orbit to turn on itself once in every 3232·5 days, about 9 years. This produces a significant lunar cycle, which the author has found of value in astro-meteorology, and would recommend to other astrologers for research.

Evection; Variation

Fig. 11 shows several principal motions or waves described by the Moon in its complicated orbit. The vertical column indicates degrees; the horizontal numbers a period of 33 days.

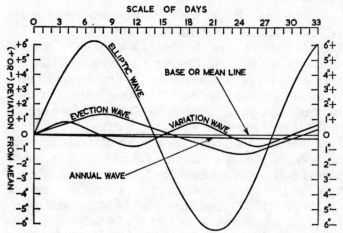

Fig. 11. Four Principal Lunar Inequalities

The *evection* (*evehere*, to carry away) is the largest lunar inequality, due to the action of the disturbing force of the Sun. This produces periodical changes in the eccentricity (shape) of the Moon's orbit, depending on the position of the line of apsides with regard to the Sun in successive lunations. This causes the Moon to oscillate about $1\frac{1}{4}°$ on each side of its mean position in a period of 31 days 19 hours, which is shown in Fig. 11 by the *evection wave*. Naturally, when the Moon is farthest from the Earth (apogee) the Sun's disturbing force will be greater than when the Moon is nearest the Earth (perigee). This inequality was discovered by Hipparchus over 2,000 years ago.

The lunar *variation* is another inequality due to the varying amount of the Sun's disturbing force. The effect is to accelerate the Moon's motion in its orbit at New Moon and Full Moon, and to retard its speed at First and Last Quarters. The period of the variation is half a lunation or 14·765 days, producing the rhythmic wave shown in Fig. 11. For astrologers the interesting feature of the *variation cycle* is that this disturbing force of the Sun varies with the 8 arcs of 45° (each 45° arc is called an *octant*, Fig. 14) which make the complete 360° cycle of the lunation, emphasizing this well-known "disturbing" astrological aspect.

The following rhythmic effects can be seen in the *variation wave* (Fig. 11): at the four quarters the effects are nil; at the first and fifth octants (midway between New Moon and First Quarter, Full Moon and Last Quarter) the Moon's motion is accelerated by about 35′; at the third and seventh octants (midway between First Quarter and Full Moon, Last Quarter and New Moon) the Moon's motion is retarded by about 35′.

Other Lunar Inequalities

At *perihelion*, when the Sun is closest, the Sun's disturbing force is greater than at *aphelion*, when it is farthest from the Earth–Moon system. The fluctuating effect of this perturbation is called the *annual equation*, which produces variations in the length of the month during the cycle of the year, and can be seen in Fig. 11 as a very gradual deviation from the Base or Mean Line. This was first recognized by Tycho Brahe in the sixteenth century.

The *parallactic inequality* is another irregularity in the Moon's motion, arising from the difference of the Sun's attraction at aphelion and perihelion. The Moon is more disturbed from Last Quarter to First Quarter than from First Quarter to Last Quarter. Since the amount of inequality depends on the distance of the "mean Moon" from the mean Sun, and on the solar parallax, the true position of the Moon is behind its mean position at First Quarter, and ahead at Last Quarter.

Because its rotation period relative to the Earth is equal to its period of "sidereal" revolution (27·32 days), the Moon always shows the same face to the Earth, hence, as viewed from the Earth, one-half is always in shadow. Actually, of course, every portion of the Moon receives sunlight in turn as it orbits the Earth. About 41 per cent of its surface is permanently hidden from us. But the other 59 per cent is visible from the Earth at some time or other due to the rotational oscillation of the Moon's disc which is explained by its three *librations* (rockings). Its axis always points in the same direction in space, is not exactly perpendicular to the plane of its orbit but inclined at an angle of 6° 5'. Thus, as the Moon passes round the Earth it inclines alternate poles toward our planet at intervals of two weeks (similar to the way in which the Earth inclines alternate poles to the Sun at intervals of six months). Alternately we see more of the Moon's north pole and more of its south pole. This oscillation is known as *libration in latitude*, and in Fig. 11 the *elliptic wave* shows this rhythmic motion.

Libration in longitude arises from the fact that, though the Moon's rotational velocity is uniform, its rate of motion round the Earth varies. At intervals of two weeks we alternately see 7° 45' "round the east side", and 7° 45' "round the west side" of its disc than we should otherwise. This does not contradict what has just been said, that the Moon's period of axial rotation and its sidereal period relative to the Earth are the same. The variation in the Moon's orbital speed is due to its increase at apogee, when it shows a little more of its *eastern* side; whereas at perigee its speed lessens and it shows a little more of its *western* side.

The *diurnal libration* is really an effect of parallax. The rotating Earth carries us from west to east, therefore when the Moon is rising we see about 57′ (Moon's mean horizontal parallax) farther round its western edge, and about 57′ round the eastern edge when it is setting.

In Fig. 11 each wave is, for convenience, shown starting at the same Base Line point. In reality this would probably never occur.

Parallax and Distance

For the astrological student interested in studying the possible varying effects of the Moon and planets in terms of their *distance* from the Earth, the *horizontal parallax* of these bodies can be obtained from *Whitaker's Almanack* for each year.

The *parallax* of a celestial body means the angle between the direction of that body as seen by the observer, and the

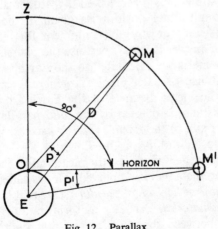

E   EARTH'S CENTRE
O   PLACE OF OBSERVATION
Z   OBSERVER'S ZENITH
M   MOON'S CENTRE
M¹   MOON'S CENTRE ON OBSERVER'S HORIZON
D   GEOCENTRIC DISTANCE OF MOON
P   DIURNAL OR GEOCENTRIC PARALLAX
P¹   HORIZONTAL PARALLAX

Fig. 12.   Parallax

direction as seen from some standard reference point. In Fig. 12 *E* represents the Earth's centre, the standard reference point; *O* is the place of observation; *M* is the Moon's centre; *M1* is the Moon's centre when on the observer's horizon. The angle from *O* to *M* to *E* is the *geocentric* or *diurnal* parallax, or simply *parallax*, of the Moon. The angle *O* to *M1* to *E* is when the Moon's centre is exactly on the observer's horizon, when the parallax is the greatest possible. The parallax is zero when the Moon (or a planet) is at *Z*, the observer's zenith, being in direct line with the observer and the Earth's centre.

The measurements *OME* or *OM1E* can be applied to any planet, but *stars* can have no geocentric parallax, since the Earth as seen from them is a mere pinpoint. For a star we speak of its *annual* parallax, meaning the angle between its direction as seen from different positions of the Earth in its annual orbit round the Sun. The position of a star on the celestial sphere is corrected for annual parallax by referring its direction to the Sun's centre; hence, *heliocentric* parallax, whose baseline is the Earth's orbit.

I have sometimes been asked by students whether a planet's longitudinal position as given in the ephemeris will be the same at any given moment for all places on Earth. The answer is "yes", because the planets' positions in the ephemeris are calculated to the Earth's centre.

The distance of each planet is generally expressed in terms of the mean distance of the Earth from the Sun, and this distance is defined by the *solar* parallax, which is the angular size of the *Earth's radius* as seen from the Sun. The equatorial radius of the Earth is used, and as its distance from the Sun varies, the mean radius (corresponding to mean distance) is employed. This is called the Sun's *mean equatorial horizontal parallax*. (*Mean*, because mean distance; *equatorial*, Earth's equatorial radius; *horizontal*, the angle between direction of the Sun on the horizon and its direction if viewed from the Earth's centre.) The mean solar parallax is 8·790″.

The *farther* away the Moon or planet is from the Earth the *smaller* its parallax. The Moon's *horizontal* parallax can be defined as the *angular semi-diameter of the Earth as seen from the Moon.* For example, when the Moon's horizontal parallax is 57′, we would also say that the Earth *as seen from the Moon* appears to have a *diameter* of 114′ (57′ ×2). The mean parallax of the Moon at apogee is 54′ 04″; at perigee, 60′ 19″. The value at mean distance is 57′ 2·63″. This corresponds to a mean distance of 60·27 times the equatorial radius of the Earth, or just over 30 Earth-diameters, or 238,857 miles.

From the details of the Moon's horizontal parallax given in *Whitaker's Almanack* for noon each day, the student can roughly assess the relative distance of the Moon from the Earth. The following table devised by the nineteenth-century astronomer, Camille Flammarian, indicates the relations which connect the angles with the distances:

Angle		Distance
1°	corresponds to	57
30′	corresponds to	114
6′	corresponds to	570
1′	corresponds to	3,438
30″	corresponds to	6,875
20″	corresponds to	10,313
10″	corresponds to	20,626
1″	corresponds to	206,265

This table can be applied to any object, celestial or otherwise. For example, if the distance of a house from where you are standing is fifty-seven times the height of the house, the height of the house will then appear equal to an angle measuring 1° of arc. If a star's parallax were 1″, its distance would be 206,265 times the Earth's mean distance from the Sun: 93,000,000 miles × 206,265!

Lunation Phases

In Chapter 6 you read that the *sidereal month* defines the Moon's period of revolution with respect to the stars; the *nodical month* defines the interval required by the Moon to orbit the Earth from one node back to the same node; the *synodic month* defines the interval between two successive New Moons. The relationship in *mean solar time* between these three lunar periods is:

$$1 \text{ synodic month} = 29 \cdot 53 \text{ days}$$
$$1 \text{ sidereal month} = 27 \cdot 32 \text{ days}$$
$$1 \text{ nodical month} = 27 \cdot 21 \text{ days}$$

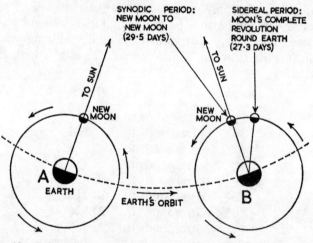

Fig. 13. Difference Between Moon's Synodic Period and Sidereal Period

Fig. 13 shows the reason for the difference between a sidereal month and a synodic month being as much as two days. For our example, let us say the sidereal and synodic months both begin at the New Moon when the

Earth is in position *A* in its orbit. When the Earth has moved to position *B* the Moon has completed exactly one synodic month, *with respect to the Sun.* Yet it would have been more than two days earlier that the position of the Moon *with respect to the stars,* as seen from the Earth, was repeated, completing a sidereal month.

The synodic month or period is more familiarly known to astrologers as a *lunation.* This cycle of the Moon begins at New Moon, when the Sun and Moon are in direct line, or *conjunction,* viewed from the Earth. The four familiar phases of the lunation are shown in Fig. 14:

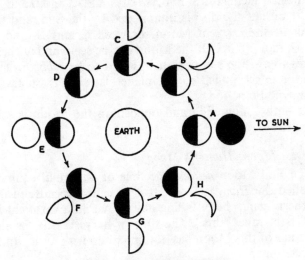

Fig. 14. The Moon's Phases

A. New Moon   ($\odot \, \text{☌} \, $☽)
C. First Quarter   ($\odot \, \square \, $☽)
E. Full Moon   ($\odot \, \text{☍} \, $☽)
G. Last Quarter   ($\odot \, \square \, $☽)

The letters *B, D, F, H* represent the intermediate points between the four principal phases. Thus, the inner partially-shaded circles show the Moon in eight successive positions, or 45° arcs called *octants*. Outside these circles, the corresponding figures show the *appearance* of the Moon as we see it in the sky at each octant. Of course, at *A* or New Moon (completely black sphere) we never see the Moon at night—contrary to the popular belief expressed by poets and songwriters!—since our satellite is then close to the Sun and at night they would both be below the horizon. At *E* the Moon appears full, because it is on the opposite side of the Earth to the Sun. At the quarters, *C* and *G*, only half the Moon's disc is visible, i.e. the half facing the Sun. Between *A* and *C*, and *G* and *A*, less than half the disc is illuminated by the Sun, and we speak of the *crescent* Moon. Between *C* and *E*, and *E* and *G*, more than half the Moon can be seen, and at these stages the Moon's appearance is said to be *gibbous*. The same term is applied to the planets at similar stages in their synodic period.

At both conjunction and opposition the Moon is said to be in *syzygy*.

Harvest Moon; Hunter's Moon

The Full Moon *nearest* the time of autumnal equinox is called the *Harvest Moon*. It is so called because in the northern hemisphere, falling always within a fortnight of 23rd September, this is the time of harvest. But the significance of this Moon has nothing to do with agricultural associations.

Whenever the Moon is in the First Point of Aries it rises from south to north of the equator. You can check this with your ephemeris, though you will find that its entry into the sign Aries may not occur in exactly the same hour that it crosses from south to north declination. Now, in northern latitudes, when declination increases north of the equator, it will be noticed that each successive day there is a corresponding increase in the Moon's length of time above the horizon. At the autumnal equinox the Sun is in ♎. Thus, when the Moon is also in or near ♈ it rises when the Sun is setting. This naturally makes the Full Moon a very conspicuous phenomenon, obviously helping the farmers with their harvesting by "lengthening the duration of daylight" as it were. On average throughout the year the Moon rises about 50 minutes later each day. This is increased when the Moon is moving south, and it decreases when it is moving north. The Harvest Moon rises around sunset for several successive evenings within *a few minutes* of the same hour.

The Harvest Moon is most marked when the ascending node of its orbit is at the First Point of Aries, for then the plane of the Moon's orbit is at its maximum inclination to the equator.

Hunter's Moon refers to the Full Moon *following* the Harvest Moon. There is a similarity, though the phenomenon is less marked. This Moon is so named since it occurs in the "hunting season".

Moon's Latitude

Raphael's Ephemerides in its "Phenomena" page (usually p. 29) gives the time when the Moon is at maximum declination, but no dates ever are listed for maximum *latitude*. It is a simple matter to find within a couple of days when the Moon is at maximum latitude by noting its daily noon position (column 7 in left-hand page for each month). As a more precise check, the Moon will be in a sign and degree that will be in *square aspect* to the position occupied by its nodes.

14 Eclipses

There are two kinds of eclipses, lunar and solar. *Eclipse* gives its name to *ecliptic*, the path of the Earth round the Sun, where this phenomenon can only occur when the centres of the Sun, Moon and Earth are in a straight, or nearly straight, line.

Lunar Eclipse

A *lunar eclipse* (Fig. 15) occurs at Full Moon ($\odot \, \beta \,)$)

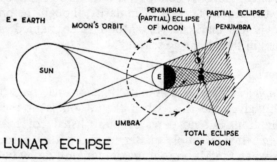

LUNAR ECLIPSE

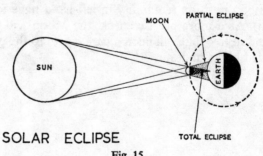

SOLAR ECLIPSE

Fig. 15

when the Earth is between the Sun and Moon, and the shadow of the Earth falls on the Moon. There are three types of lunar eclipse: *total*, *partial*, and *penumbral*.

A *total lunar eclipse* occurs when the Moon enters the *umbra*, the direct conical-shaped shadow of the Earth, and no portion of the Moon's disc receives direct light from the Sun. The phenomena begins gradually as the Moon enters the *penumbra*, a region from which part of the Sun's surface would be seen (from the Moon). The bright glow of the Full Moon's Sun-reflecting surface is increasingly dimmed, until on its entering the *umbra*, part of the Moon is completely darkened, partially eclipsed. The duration of totality (whilst the Moon is completely within the umbra region of the Earth's shadow) averages 1h. 40m. The time from the first contact or appearance of the umbra on the disc, until the Moon has finally left the umbra, is almost 4 hours.

A *partial lunar eclipse* occurs when the Moon's latitude is sufficiently greater at the time of $\odot \, \mathcal{S} \, \mathbb{D}$ than it must be for a total eclipse, so that its disc is never wholly covered by the umbra.

A *penumbral lunar eclipse* occurs when the Moon's latitude is such that the Moon only passes through the penumbra region of the Earth's shadow.

Solar Eclipse

A *solar eclipse* (Fig. 15) occurs at New Moon ($\odot \, \mathcal{d} \, \mathbb{D}$) when the Moon passes between the Sun and Earth. The eclipse can be of three types: *total*, *partial*, and *annular*.

The phases of a *total solar eclipse* are similar to those of a total lunar eclipse. However, since the Earth's size is greater than that of the Moon, only parts of the Earth's surface are eclipsed. Therefore, as the Sun and Moon approach conjunction the "track of the eclipse" passes across a narrow section of the hemisphere facing the Sun and Moon, first as the penumbral shadow and then the almost complete darkness within the umbra. The duration of a total eclipse is greatest at the equator, lasting as much as $7\frac{2}{3}$ minutes. Around latitude 45° the duration can be as much as $6\frac{1}{2}$ minutes. The time from the first contact of the edge of the Moon with the Sun's disc, until the Moon finally "clears" the Sun, may be a little over 2 hours.

The *partial solar eclipse* occurs when the Moon's disc does not totally cover the Sun. This may be seen at the time of a total eclipse, by observers outside the track of the Moon's *umbra*, who are within the area of the *penumbra*.

The necessary condition for a solar eclipse to be *total* is that the measurement of the Moon's semi-diameter must *exceed* the Sun's semi-diameter (as seen from the Earth). The Sun's semi-diameter varies (according to the Earth's distance from the Sun) between 15·8′ and 16·3′, while the Moon's semi-diameter varies between 14·7′ and 16·8.′ Therefore if the Moon is less than the Sun's minimum (15·8′) a solar eclipse cannot be total. It will be either partial or *annular*. If the Moon's semi-diameter is greater than the Sun's maximum (16·3′) a solar eclipse cannot be annular, and is either total or partial. A *total* eclipse occurs, therefore, when the Moon is at or near *perigee*. An *annular* eclipse occurs at or near Moon's *apogee* when the Moon is so far from the Earth that at maximum obscuration its disc does not wholly cover the Sun, but leaves a ring of light still visible. Hence the name of this type of eclipse (Latin: *anulus*, a ring).

Ecliptic Limits

Since eclipses occur only when the centres of the Sun, Moon and Earth are in a straight, or nearly straight, line, it follows that both the Sun and Moon must be at or near one of the *Moon's nodes* at New or Full Moon. Thus, there are certain limits conditioning eclipses, called *major* and *minor ecliptic limits*.

Major ecliptic limit: a lunar eclipse *may* occur if the

Full Moon is within 11° 38′ (longitude) of a node; a solar eclipse *may* occur if the New Moon is within 17° 25′ (longitude) of a node.

Minor ecliptic limit: a lunar eclipse *must* occur if the Full Moon is within 9° 39′ of a node; a solar eclipse *must* occur if the New Moon is within 15° 23′ of a node. If New or Full Moon occur within 5° of a node the eclipse will be total.

In terms of the Moon's *latitude*, a lunar eclipse cannot occur unless at Full Moon the Moon's latitude is less than about 56′. Its latitude must not exceed 26′ for a *total* lunar eclipse to occur. At New Moon a solar eclipse cannot occur unless the Moon's latitude is less than 1° 28′, and less than 58′ for a *total* solar eclipse or for an *annular* eclipse.

The period of revolution of the Moon's nodes, relative to the Sun, marks the recurrence of the intervals of time during which eclipses can occur, since eclipses are only possible when the Sun is near to a node. This period is known as the period of a *synodic revolution of the nodes*. Thus, eclipses of a given year always occur at two opposite seasons, near the time when the Sun crosses the nodes of the Moon's orbit.

The *greatest* number of eclipses possible in one year is seven. Five solar and two lunar, or four solar and three lunar. The *least* number possible in one year is two: both solar.

The Saros of the Chaldeans

Since the Moon's nodes move westward around the ecliptic once in about 19 years the time taken by the Sun to pass from a node to the same node again (synodic revolution of the nodes) is about 346·62 days. This is sometimes called the *eclipse year*. The Chaldean astronomers discovered that after 6585·78 days, or 18 years 11⅓ days (or 10⅓ days if there have been five leap years in the interval), the Moon's nodes will have performed 19 revolutions relative to the Sun (i.e. 19 eclipse years). 223 lunar or synodic months also equal 6585·32 days. Therefore the Sun and Moon will return almost exactly to the same position relative to the nodes at the end of this cycle, and the Chaldeans called this cycle the *Saros*. The average number of eclipses in a Saros is seventy.

Metonic Cycle

A lunar cycle introduced into Greece by Meton and Euctemon in 432 B.C., though it was already known in eastern countries. After a period of 19 years, New and Full Moons recur on the same days of the year. The Metonic Cycle is 235 synodic (lunar) months, or 6939·69 days in length; almost exactly 19 tropical years.

15 The Seasons

Cause of the Seasons

The revolution of the Earth in its orbit, and the resulting changes in the direction of its axis of rotation relative to the Sun, produces the seasons.

Hence, the varying angle at which the Sun's rays strike different parts of the Earth's surface is the principal cause of the seasons. The higher the Sun's meridian (noon) altitude, that is, the closer the Sun is to being directly overhead, at a given place, the greater the amount of heat that will be received. It follows that the higher the Sun's altitude is at noon for a place, the longer the Sun will be above the horizon during the day for it to attain the greater altitude. Therefore, it is not only the height of the Sun in the sky at noon that affects temperature, but also the length of the interval from sunrise to sunset.

In Fig. 16 we see the changing direction of tilt of the Earth's axis relative to the Sun during the cycle of the year. The axis always points to the Pole Star, but the

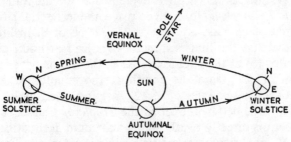

Fig. 16. Inclination of the Earth's Axis, the Main Cause of the Seasons

changing position of the Earth relative to the Sun causes this angle of tilt to bring a greater portion of the Northern Hemisphere more directly "under the Sun's rays" at the summer solstice (21st June) than at the winter solstice (21st December), when a greater portion of the Southern Hemisphere experiences the Sun more directly overhead. At the equinoxes (21st March and 21st September) equal areas of both hemispheres are exposed to equal amounts of the Sun's rays, when day and night are of equal length.

Effects of the Atmosphere

The Earth's atmosphere tends to hold back some of the Sun's heat. In the Northern Hemisphere the normally hottest time of the year is, therefore, not when the days are longest and the Sun's rays most vertical (21st June), but about a month later, due to the accumulation of heat absorbed by the atmosphere. Likewise, the normally coldest time of the year is nearer 1st February than 1st December (winter solstice).

Earth Zones

The derivation of the five well-known zones into which the Earth sphere is divided is illustrated in Fig. 17.

As the Sun's declination (actually the angle of tilt of the Earth's axis relative to the Sun) ranges between 23° 27′ N. and 23° 27′ S., it is evident that for the Sun to be *vertically overhead* for a given place, the terrestrial latitude of that place must not exceed 23° 27′ N. or S. In Fig. 17 we see that the Sun is directly overhead at noon at the equator (*b*) at the equinoxes. Also, the Sun is never more than 23° 27′ from the zenith at the equator at either solstice. These angles of 23° 27′ are indicated by latitude 23° 27′ N., better known as the *Tropic of Cancer*, where the Sun is directly overhead, at maximum declination 23° 27′ N., at the summer solstice, when it enters the sign ♋; and latitude 23° 27′ S., or the *Tropic of Capricorn*, where the Sun is directly overhead, at maximum declination 23° 27′ S., at the winter solstice, and enters the sign ♑.

The region between the tropics is called the *Torrid Zone*.

It is seen that the inclination of the Earth's axis produces a difference in the duration of day and night, so that the greater the latitude of a place above or below the equator,

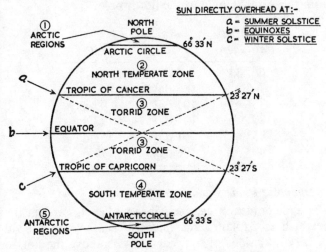

Fig. 17. Derivation of the 5 Earth Zones

the greater is the length of the longest day for that place, and the shorter will be its shortest day. When we reach a distance from the north pole equal to the inclination of the Earth's axis (90° − 23° 27′ = latitude 66° 33′ N.) the Sun does not set on the day of the summer solstice. It remains above the horizon for 24 hours. Thus, latitude 66° 33′ N. defines the *Arctic Circle*.

Similarly, 66° 33′ S. defines the *Antarctic Circle*, where at the winter solstice the Sun remains above the horizon for about 24 hours, and is known as the *midnight Sun*.

At the poles the Sun never rises more than 23° 27′ above

the horizon, hence the icy nature of the *Arctic* and *Antarctic Regions.* Also, the poles experience alternately 6 months of what are called *Perpetual Day* and 6 months of *Perpetual Night.*

In the following table,* *A* is the approximate period during which the Sun never sets, *B* is the approximate period during which the Sun never rises, for latitudes from the Arctic Circle to the north pole:

Northern Latitudes	A	B
66° 33'	1 day	1 day
70°	65 days	60 days
75°	103 days	97 days
80°	134 days	127 days
85°	161 days	153 days
90°	186 days	179 days

Circumpolar Planets and Stars

The phenomenon just described makes this an apt time in which to explain *circumpolar planets.* When the Sun for certain periods of the year remains above the horizon at a given latitude, it is then a *circumpolar star.* Likewise, a circumpolar planet is a planet whose angular distance from the pole is *less than* the polar elevation of a given place. The planet will transit the meridian of the place twice in 24 hours, once *above the pole,* and once *below the pole,* but throughout the period it will remain *above the horizon* (Fig. 18). Any planet, including the Moon, can be circumpolar when its declination is near to that of the Sun's, and it is in or near the sign containing the Sun, or opposite to that of the Sun, at the time of the Sun being (or nearly being) circumpolar.

This book is not intended to cover the explanations of

* From *Popular Astronomy* by Camille Flammarion.

house systems in astrology, but the author would take this opportunity of stressing the utter fallacy of the still commonly used Placidean House System. When Placidean house cusps are calculated for birthplaces in higher latitudes the houses become grossly distorted, ridiculously out of proportion to the twelve-fold theory or purpose of

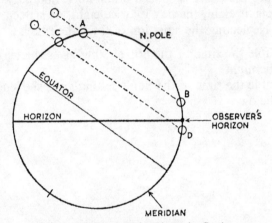

Fig. 18.   Circumpolar Planet (or Star)
Planet or star transiting meridian at *A* and *B* is circumpolar
(remains above horizon)
Planet or star transiting meridian at *C* and *D* is *not* circumpolar
(passes below horizon)

these divisions of the chart. But worse than this, at certain periods whole areas of the ecliptic degrees can never form house cusps or even be included in the houses, and if any planets or the Sun and Moon (being circumpolar) are within these degree-areas they too have to be omitted from the chart! If a house system is not valid for a birth *anywhere on Earth* then it should be discarded.

Sunrise and Sunset

Because at apparent noon and midnight the Sun is midway between the eastern and western parts of the horizon, it follows that the time taken from sunrise to noon, and noon to sunset, are equal. There is a simple method by which we can learn the length of the day and night, given the times of sunrise and sunset. For example, if the Sun rises at eight, the time from midnight to sunrise is eight hours; which equals the time from sunset to midnight: therefore the night is sixteen hours long. Similarly, if the Sun sets at eight, the day is twice eight, or sixteen hours long. Neglecting the equation of time, the rule is:

Double the time of the Sun's rising gives the length of the night.

Double the time of the Sun's setting gives the length of the day.

16 *The Tides*

The rotation of the Earth and the combined attraction of the Sun and the Moon on the waters of the Earth cause their periodic ebbing and flowing which we know as the *tides*.

The Moon being nearer to the Earth than the Sun is to the Earth, its *disturbing force* or *tide-generating force* is about two-fifths greater than that of the Sun. That the Moon's effects are greater is evident when the Moon is at *perigee* (closest point in its orbit to the Earth), when the tides are nearly 20 per cent higher than when the Moon is in *apogee* (farthest point); also, on average, there are two high and two low tides in 24h. 50m., which is precisely the same average interval between two successive passages of the Moon across the meridian.

Flood tide is while the water is rising; *ebb* tide is while it is falling. *High-water* is the moment when the water-level is highest; *low-water* when it is lowest.

Each month the Sun–Moon tide-raising power is greatest at *New* or *Full Moon*, and these tides are known as *Spring Tides*. Both bodies are in line relative to the centre of the Earth and the combined disturbing force produces a pull along the line AA^1 (Fig. 19), and a "squeeze" along the line BB^1. At these times the interval between the corresponding tides of successive days is less, at 24h. 38m., than the average, 24h. 50m. Spring tides are the highest possible when the Sun and Moon are both in *perigee* (about January), the tide-raising force being about 30 per cent larger than at *apogee*.

Tides occurring at the Moon's First and Last Quarter

are called *Neap Tides*, when the solar and lunar disturbing forces are at right angles (Fig. 19). The height of the neap tide is the difference of the heights of the separately generated lunar and solar tides, which is less than half as great as the spring tides. Neap tides are the most marked when the Moon is in *apogee* but the Sun is in *perigee* (around January). The Sun then pulls *against* the Moon with its greatest power ratio so far as their combined tide-disturbing action upon the Earth is concerned.

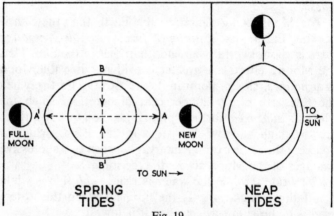

Fig. 19

Between New Moon and First Quarter the interplay of the solar and lunar disturbing forces displace the position of high-water to the west, and high-water occurs *earlier* than it would if affected only by the Moon. The tides are then said to *prime*.

Between First Quarter and Full Moon the solar and lunar interplay displaces the high tides to the east, and these occur *later* than they would if due to the Moon's influence alone. The tides are then said to *lag*.

Similarly, between Full Moon and Last Quarter the tide *primes*, and between Last Quarter and New Moon the tide again *lags*.

17 *Planetary Aspects*

This chapter is not intended to explain all the commonly used planetary aspects in astrology, but simply to describe the planets' positions in the sky *as viewed from the Earth* at times of major configurations formed relative to the Sun.

An *aspect* or *configuration* is the angular distance measured along the ecliptic in celestial longitude, as viewed from the Earth.

The outside circle in Fig. 20 represents the orbit of a *superior* planet (a planet outside the Earth's orbit). The inner circle, closest to the Sun, the orbit of an *inferior* planet, Mercury or Venus.

Conjunction

When a superior planet is in line with the Sun and Earth, and is on the far side of the Sun, it is in *superior conjunction*, or just simply *conjunction*. When an inferior planet is similarly positioned on the far side of the Sun from the Earth it is in *superior conjunction* also; but when it is on the near side of the Sun and in line with that body and the Earth it is in *inferior conjunction*. All planets can reach superior conjunction, but only Mercury and Venus form the inferior conjunction. When a planet or the Moon is in conjunction with the Sun it is always on or near the meridian *at noon* for any place. The interval between successive superior conjunctions, or successive inferior conjunctions, is known as a planet's synodic period (p. 58).

Elongation

The apparent angular distance of a planet east or west

from its centre of motion (the Sun) at any time is called its *elongation*. With superior planets elongation can be measured up to 180° east or west from the zero point at conjunction. The expression "at elongation" usually means the planet's maximum angular distance during its particular revolution of the Sun. But in the case of the inferior planets, Mercury can only be just over 28° from the Sun (viewed from the Earth) at *Greatest Elongation East* or *Greatest Elongation West*; Venus can only reach a distance of about 48° at these angular limits. It follows, therefore, that when a planet is in conjunction the Sun elongation is nil; at opposition, elongation is 180°.

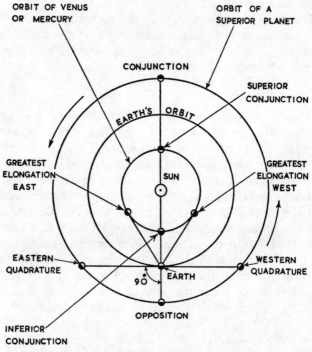

Fig. 20. Planetary Aspects and Elongations

Fig. 20 shows an inferior planet's path to be through superior conjunction to Greatest Elongation East, then swinging westwards through inferior conjunction until it reaches its greatest western elongation, and so back again to superior conjunction to complete a synodic period.

When Mercury and Venus are east of the Sun they rise or set later than the Sun, because even though in terms of celestial longitude they will then be ahead of the Sun, moving in a direct or *anti-clockwise* direction (as direction of arrow, Fig. 20), the rotation of the Earth makes the Sun and planets appear to move in a *clockwise* direction.

Thus, when their Greatest Elongation East occurs in the spring months, when they are *north* of the celestial equator, they will be favourably placed for observation as "evening stars", being high in the sky at sunset. Venus, in particular, is then very bright and sets about 4½ hours after sunset, and Mercury about 2 hours after sunset. Mercury is only observed by the naked eye about the time of greatest apparent elongation. In the autumn months, when *south* of the celestial equator, neither planet attains a very high altitude, and when western elongation occurs at this time of the year they set before darkness has completely fallen, setting before the Sun. At western elongation Mercury and Venus are "morning stars" because they rise ahead of the Sun, before daylight.

Quadrature

Only the superior planets, which have orbits larger than the Earth's, can form square (quadrature) and opposition aspects with the Sun. The Moon, of course, because it orbits the Earth also forms these aspects. To an astrologer, when a planet's angular distance along the ecliptic from the Sun is 90° it is in *square* aspect; but the astronomer speaks of the planet being at *quadrature* (Fig. 20). A planet is at eastern quadrature when, viewed from the Earth, it is 90° eastwards of the Sun, at apparent midpoint between conjunction and opposition. This would be First Quarter in the case of the Moon. At western quadrature a planet is westwards of the Sun by 90° at apparent midpoint between opposition and conjunction, and for the Moon this would be Last Quarter.

At eastern quadrature a superior planet's meridian passage is around 6 p.m.; at western quadrature, 6 a.m.

Opposition

A planet is in *opposition* to the Sun when the Sun, Earth, and planet are approximately or directly in line, with the Earth in the middle. Mercury and Venus cannot be in opposition to the Sun. When at opposition a superior planet (or the Moon) crosses the meridian at about midnight. A planet is also at its closest point in its orbit to the Earth (perigee) at opposition, and at its farthest point (apogee) at conjunction the Sun, as shown in Fig. 20. This means that at opposition it will be above the horizon all night.

Occultation

Occultation (Latin: *occulere*, cover) is a term most commonly applied to the hiding of a star or planet by the Moon. In other words, the other body is "eclipsed", and it would not be incorrect to say that a total solar eclipse is an occultation of the Sun. An occultation can only occur when the other body and the Moon are in the same degrees of longitude and of declination.

18 *Retrograde Motion*

All planets revolve around the Sun from west to east, which is termed their *direct* motion. Due, however, to the Earth's own orbital motion in the same direction, the other planets at certain times appear to the observer on the Earth to gradually slow down until they "stop" or "stand still", before moving in the reverse (clockwise) direction to normal.

When a planet moves in this reverse direction along the ecliptic it is said to be *retrograde*. This is indicated in the ephemeris by the capital letter "R" against its longitudinal position. When a planet appears to be motionless, before turning retrograde, or before changing from retrograde to direct motion again, it is at its *stationary points* or *stations* and it is then said to be *stationary*.

As an example we can turn to page 13 in *Raphael's 1965 Ephemeris*. On 28th June Saturn, as indicated by "R", becomes stationary at ✕ 17° 13'. In fact, on page 33, for 28th June we see that this occurs exactly at 5.30 a.m. E.T. Saturn then begins to retrograde, less than 1' each day to begin with, until during August its daily retrograde motion reaches 5'. It is not until 3.28 a.m. on 14th November 1965 that Saturn is again stationary and begins its direct west-to-east orbital motion again, as indicated by the capital "D".

The Sun, being the axis for the Earth's orbital motion, can never appear retrograde. Neither can the Moon.

Superior Planets

In Fig. 21 retrograde motion is illustrated in two diagrams. The top diagram shows the apparent "loop" performed by a superior planet in the process of reaching two stationary points which enclose between them the arc of retrogression (or retrogradation). Between the arc *ABC* the planet appears to move *eastwards* (direct) with reference to the stars or zodiac, its angular motion decreasing as it approaches *C* (stationary point). Along the arc *CDE* the planet appears to move *westwards* (retrograde) until *E* (stationary point) is reached. The arc *EFG* traces its eastward and direct motion again. As can be deduced from the positions of Sun, Earth, and superior planet in the lower diagram, retrograde motion occurs only around the time of opposition for a superior planet.

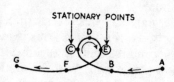

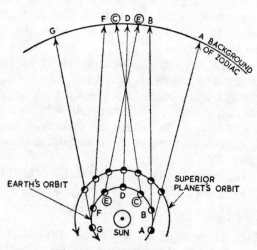

Fig. 21. Retrograde Motion

Inferior Planets

With Venus the period of retrogression always occurs about 3 weeks before inferior conjunction and extends for 6 weeks, direct motion beginning again about 3 weeks after inferior conjunction. Venus therefore has to be between the Sun and Earth for their angular relationship to produce apparent retrograde motion. In a synodic period (successive superior conjunctions with the Sun) of 584 days, Venus is retrograde for an average of only 42 days, as against direct motion for 542 successive days.

Mercury, like Venus, can only be retrograde when nearest to the Earth, between the Earth and the Sun. Midway between Greatest Elongation East and inferior conjunction Mercury appears to be stationary because (as with Venus similarly placed) for a short time it is travelling exactly towards the Earth. An inferior planet is, therefore, *always* retrograde at inferior conjunction. Mercury's period of retrogression averages 20–24 days, when the planet again appears stationary, midway between inferior conjunction and Greatest Elongation West, due to its then travelling exactly away from the Earth. In a synodic period of 116 days Mercury retrogrades for an average of 22 days, and moves direct for 94 days. Mercury is retrograde for about 20 per cent of its synodic period, Venus for only 7 per cent.

Fig. 21 can be used to illustrate the retrograde motions of Mercury and Venus if we think of the orbit shown for the superior planet as the orbit of the Earth and the orbit that is indicated as the Earth's as that of either of the two inferior planets. In this sense we see that, viewed from the superior planet towards the inferior planet, retrogression does actually occur between midway point Greatest Elongation East–inferior conjunction and inferior conjunction–Greatest Elongation West (*B* and *F* representing maximum elongation). Similarly, as viewed from Mars, the Earth would always appear retrograde around its inferior conjunction with the Sun.

19 *The Rotating Signs relative to the Houses*

A not unnatural puzzle to beginners in astrology is *why do the signs rotate in the opposite direction to the numbering of the houses?*

As we know, the houses in the astrological chart are numbered *anti-clockwise* from 1 to 12, beginning on the left-hand (east) side of the chart. The zodiacal signs, beginning with the first sign, Aries, through to the twelfth sign, Pisces, are also listed *anti-clockwise*.

The beginner's problem starts when he learns that a planet rising over the Ascendant (east point of chart) must move in a *clockwise* direction. And yet this same planet by progression and moving direct will be plotted against the natal chart in an *anti-clockwise* direction! It seems perfectly logical, therefore, that planets and signs, instead of rising and setting in a *clockwise* direction, should move in the same direction as the numbering of the houses, *anti-clockwise*. Or, better still, thinks the puzzling student, the houses should be numbered *clockwise*.

Here is the explanation. It is perfectly correct for the houses of the chart to be numbered in an anti-clockwise direction. In reality the planets move in this same direction, as is evident by their progressed direct motion—which means that this anti-clockwise direction is also proper for the signs. It is because the Earth, which is also rotating on its axis, *moves at a much greater speed than the planets moving in their orbits* that the planets and signs *appear to be* moving in the opposite (clockwise) direction. If the

Earth were motionless, fixed in space, neither rotating on its axis nor orbiting the Sun, then we would see the Moon moving in an anti-clockwise arc over our heads, and be able to trace at much slower rates this same direction of the planets' motions.

20 *Additional Definitions*

Angles, the

This is not a term used by astronomers, but by astrologers with regard to the extremely important great circles of reference: the *Ascendant–Descendant* points of intersection of ecliptic–horizon, and the *M.C.–I.C.* points of intersection of ecliptic–meridian.

Arc

Part of the circumference of a circle. For instance, distance measured along the ecliptic between the M.C. and Ascendant would be measurement along an *arc* of the ecliptic circle defined by those two points.

Colures

The declination circle (meridian of longitude) passing through the north and south poles of the equator and the equinoctial points is called the *Equinoctial Colure*. The declination circle passing through the poles of the equator and the solstitial points is called the *Solstitial Colure*.

Heliacal Rising or Setting

The rising or setting of a star at the same time as the Sun is called its *heliacal* rising or setting. The Babylonians, nearly 3,000 years before the Christian era, recorded the heliacal rising of stars.

Ingresses

Astronomically *ingress* refers to the "entry" of a body or planet into the disc of another body, such as the transit of Venus across the face of the Sun, at the moment the two discs meet. The passing of one body off the disc of another is called its *egress*. In astrology *ingress* describes the entry of a celestial body into a sign, most commonly referred to as the Sun's entry into Aries (Vernal Ingress), Cancer (Summer Ingress), Libra (Autumnal Ingress), and Capricorn (Winter Ingress).

Light-Year

The distances of the stars is so enormous that a unit called the *light-year* is used to express these distances. It represents the distance travelled by light in one year, which is about 63,000 times the distance of the Earth from the Sun. The velocity of light is 186,285 miles a second. For instance, instead of saying the star Castor is 304 billion miles distant, one could say it is about 52 light-years away. The light by which Castor is seen left that star about 52 years previously!

An even larger and more commonly used unit in technical astronomical work is the *parsec*. This is the distance corresponding to a parallax (*par-*) of one second (*-sec*) of arc. A star with a parallax of 1″ is at a distance of 3·26 light-years or 1 parsec. For even greater distances the *kiloparsec* (a thousand parsecs) and the *megaparsec* (a million parsecs), or alternatively the *kilolight-year* or *megalight-year*, are employed as units. The diameter of our galaxy is believed to be about 25 kiloparsecs (25,000 parsecs) across; which is to say that light would take around 81,500 years (25,000 × 3·26) to travel from one side of the galaxy to the other. When we think that light travels at over 186,000 miles a second the length of our galaxy's diameter in actual miles reaches a fantastic proportion! Light from our Sun takes 499 seconds, or 8 minutes 19 seconds, to reach us; from the Moon, about $1\frac{1}{2}$ seconds; from the outermost planet, Pluto, $5\frac{1}{2}$ hours; from the most distant bodies yet photographed, 500 million years (500 megalight-years).

Lune

A portion of a sphere contained between two great semi-circles. Each of the twelve sections of a sphere representing the twelve houses in astrology is a lune.

Nonagesimal

The highest point of the ecliptic above the horizon. Therefore exactly 90° from both Ascendant and Descendant. Not to be confused with the Midheaven (M.C.) which is the point where the ecliptic intersects the upper meridian. Only rarely will the M.C. coincide exactly with the nonagesimal, which can be seen should this occur if one sets up a chart by the Equal House System. By this system of house-division the twelve houses are of equal 30° length, therefore the nonagesimal will always be the tenth house cusp. Usually the M.C. will be seen to fall several or many degrees either side of the nonagesimal. The M.C. and the nonagesimal refer to the ecliptic, but the zenith, which might also be confused with either of the other two points, does not refer to the ecliptic but to a point directly above the observer. The zenith cannot be plotted in the astrological chart. Only charts calculated for births on the equator would be likely to occur when the M.C., nonagesimal, and zenith might occupy the same point in "the middle of the heavens".

Quadrant

The fourth part of the circumference of a circle; an arc, the angle of which is 90°.

Signs of Long and Short Ascension

Due to the ecliptic lying at an oblique angle across the equator (Fig. 5) certain signs in mid-latitudes take much longer to rise above the horizon than do others. In the Northern Hemisphere (mid-latitudes) the signs Cancer to

Sagittarius are referred to as signs of *long ascension*; those from Capricorn to Gemini are called signs of *short ascension*. The opposite applies for mid-latitudes in the Southern Hemisphere. As the name suggests, signs of *short* ascension rise quickly. You can check this yourself in a Tables of Houses. Around latitude 50° N. the signs Pisces and Aries take a mere 52 minutes or so for their complete 30° to rise over the horizon. Whereas Virgo and Libra take approximately 2 hours 50 minutes.

These terms do not apply in equatorial regions, whilst in polar regions some signs do not rise at all, but are circumpolar (*q.v.*).

Southing

The moment a celestial body crosses the upper meridian of a place it is *southing* that meridian. *Whitaker's Almanack* and the *Nautical Almanac*, for instance, give the times when the planets *south* the Greenwich meridian. The Sun always souths a meridian around noon. The name derives from the fact that this transit occurs at the *south point* of the horizon. To understand this, and why south is shown at the *top* of the astrological chart and north at the *bottom*, we have to think of ourselves as the observer standing facing south. East (Ascendant) is on our left where the Sun rises over the horizon. In correct sequence the Sun moves in an arc to the south (high in the sky), then dips below the horizon in the west (Descendant) on our right. Around midnight it reaches the north point of the chart below the horizon if we plot its motion.

Transit

The passage of a planet (or star) across the upper or lower meridians is called its *transit*. In astrology the term transit is used more broadly, referring to a conjunction or other aspect to any important factor in a chart by an *actual* position of a planet at a given time, past, present, or future—as distinct from *progressed* planets and symbolic measures.

Tropical Zodiac, Sidereal Zodiac

The *Tropical Zodiac* refers to the astrological system based on the calculated motions of the Sun and planets "through" the Signs of the Zodiac, and is the most commonly used by western astrologers. The signs have a logical and calculable starting-point: the intersection of ecliptic and equator. The *Sidereal Zodiac* refers to the positions of the Sun and planets as seen against the actual constellations. The starting-point for the first degree of Aries cannot possibly be reliable. The constellations are simply groups of stars of immensely varying sizes.

Index